DATE DUE

SEP 15 1992			
		SEP 15 1992	
JAN 15 1993	BIRD		
MAY 15 1993		MAR 23 1993	
MAY 31 1994	ENDS		
		MAY 23 1994	

SETTLEMENT SCHEMES IN
TROPICAL AFRICA

SETTLEMENT SCHEMES IN TROPICAL AFRICA

A study of organizations and development

ROBERT CHAMBERS

FREDERICK A. PRAEGER, *Publishers*

New York · Washington

BOOKS THAT MATTER

Published in the United States of America in 1969
by Frederick A. Praeger, Inc., Publishers
111 Fourth Avenue, New York, N.Y. 10003

Library of Congress Catalog Card Number: 76-86248

Printed in Great Britain

To my Parents

Contents

4 Changing the Land

5 Ruling the Settlers

6 Organizing Production and Growth, 1960-1966

PART III SETTLEMENT SCHEMES AS ORGANIZATIONS

7 The Actors: Staff and Settlers

8 Staff-Settler Organization

Tables and Charts

Maps

Plates

All photographs are of the Mwea Irrigation Settlement.

Abbreviations and Conventions

1 *Short titles for schemes*

The full title for each scheme is given the first time it is mentioned in the text. Subsequently, most schemes are referred to in an abbreviated form, as they often are colloquially. The most common abbreviations used are:

Short title	Full title and country
Ahero	Ahero Pilot Irrigation Project, Kenya
ALDEV Schemes	African Land Development Organization Settlement Schemes, Kenya
Anchau	Anchau Rural Development and Settlement Scheme, Nigeria
Chesa	Chesa Native Purchase Area Land Settlement Scheme, Rhodesia
Damongo	Gonja Development Company Scheme, Damongo, Ghana
Daudawa	Daudawa Land Settlement Scheme, Nigeria
Eastern Nigerian Settlements	Settlements of the Eastern Nigerian Farm Settlement Programme
Fra Fra	Fra Fra Resettlement Scheme, Ghana
Gezira	Gezira Scheme, Sudan
Ilora	Ilora Farm Settlement, Western Nigeria
Kariba	Kariba Resettlements, Rhodesia and Zambia
Kenya Million-Acre Scheme	Million-Acre Settlement Scheme, Kenya
Khasm-el-Girba	Khasm-el-Girba Project, Sudan
Kigezi	Kigezi Resettlement Schemes, Uganda
Kigumba	Kigumba Settlement Scheme, Uganda
Kiwere	Kiwere Settlement, Tanzania
Kongwa	Kongwa Farming Settlement Scheme, Tanzania
Managil	Managil South-Western Extension of the Gezira Scheme, Sudan
Mubuku	Mubuku Irrigation and Pilot Demonstration Project, Uganda
Muhuroni	Muhuroni Complex Sugar Scheme, Kenya
Mungwi	Mungwi Settlement Scheme, Zambia
Mwea	Mwea Irrigation Settlement, formerly the Mwea-Tebere Irrigation Scheme, Kenya
Nachingwea	Nachingwea Tenant Farming Scheme, Tanzania

Niger Agricultural Project	Niger Agricultural Project at Mokwa, Nigeria
Nyakashaka	Nyakashaka Resettlement Scheme, Uganda
Ol Kalou	Ol Kalou Salient, Kenya
Orin-Ekiti	Orin-Ekiti Farm Settlement, Western Nigeria
Perkerra	Perkerra Irrigation Scheme, Kenya
Rwamkoma	Rwamkoma Pilot Village Settlement, Tanzania
Sabi	Sabi Valley Irrigation Projects, Rhodesia
Shendam	Shendam Resettlement Scheme, Nigeria
South Busoga	(second) South Busoga Resettlement Scheme, Uganda
Sukuma	Sukumaland Development Scheme, Tanzania
Tanzania Pilot Village Settlements	Pilot Village Settlement Programme, Tanzania
Tema	Tema-Manhean Resettlement, Ghana
Uganda Group Farms	Group Farming Programme, Uganda
Upper Kitete	Upper Kitete Pilot Village Settlement, Tanzania
Urambo	Urambo Tobacco Scheme, Tanzania
Volta	Volta River Resettlements, Ghana
Western Nigerian Settlements	Settlements of the Western Nigerian Farm Settlement Programme
Zande	Zande Scheme, Sudan

2 *Other Abbreviations*

A.A.O.	Assistant Agricultural Officer
A.D.C.	African District Council
ALDEV	African Land Development Organization
D.A.O.	District Agricultural Officer
D.C.	District Commissioner
D.O.	District Officer
J.I.C.	Joint Irrigation Committee
L.N.C.	Local Native Council
Local Committee	The Mwea-Tebere Irrigation Scheme Committee, later the Mwea Irrigation Settlement Committee
M.I.S. Annual Report	Mwea Irrigation Settlement Annual Report
M.O.W.	Ministry of Works (used also to refer to its predecessor the Public Works Department)
P.A.O.	Provincial Agricultural Officer
P.C.	Provincial Commissioner

3 *Conventions*

The term 'Agricola' is used to denote professional agriculturalists and agricultural managers, regardless of their departments. It can thus, in the description of the history of the Mwea Irrigation Settlement, be used for officers of the Department of Agriculture and of ALDEV. 'The Agricolas', 'the Engineers', and 'the Administration' are used with equal weight in discussions of departmental and professional attitudes to describe agriculturalists (Department of Agriculture and ALDEV), hydraulic engineers (M.O.W.), and administrative officers (Provincial Administration), respectively.

To avoid overburdening the text with footnotes only the more important sources are cited. For some sections, however, sources are given in more detail in my thesis: 'The Organization of Settlement Schemes: a comparative study of some settlement schemes in anglophone Africa, with special reference to the Mwea Irrigation Settlement, Kenya', Manchester University, 1967. The present book, however, incorporates major revisions and additions so that a source may not always be traceable.

Preface

This book originates in personal, somewhat amateurish involvement in rural development projects. While serving in the Administration in Kenya I had some responsibility for two schemes, both of which had chequered histories not unrelated to the way in which they were managed. The first was a grazing control programme which subsequently collapsed, and the second a water reticulation scheme which ran into financial difficulties. This doubtful practical record qualified me for transfer to the safer task of teaching administration, which provided incentive and opportunity to study the organization of development schemes and the behaviour of those who initiate and manage them. At first the Perkerra Irrigation Scheme in Kenya seemed a suitable case for examination, but later I changed to the Mwea Irrigation Settlement which had better written records and a less unusual history.

At the same time social scientists in East Africa were beginning to take increased interest in settlement schemes. Following discussions with some of them, and a steady accumulation of studies from Central and West Africa, it began to appear that settlement schemes might provide a coherent and intelligible field of study. The attention being given by governments to plans for future settlement suggested that work in this field might have some practical use, and the absence of any body of ideas that could be called a theory of settlement promised the pleasures of exploration. I therefore visited what schemes I could, and was fortunate in being able to spend six weeks in Ghana studying the Volta River Resettlements, as well as visiting settlement schemes in Kenya, Uganda, Tanzania, and, fleetingly, the Sudan. Gradually, the study came to centre on settlement schemes as organizations.

The deeper I went into the history of the Mwea Scheme, the more I came to see that interdepartmental relations were significant. In addition, my experience working in government had made me feel that these relations were important in understanding how and why government acted as it did. At the same time, it was evident that they were a neglected aspect of development schemes in Africa, both in description and in planning. It therefore seemed useful to pay attention to them and in particular to organizational conflict. The result, however, is not intended to be, nor is it, an attack on colonial or any

xix

other government, nor on any profession, department or person. Those who work in bureaucracies, universities not least, know that departmentalism is more or less universal, and that attempts are sometimes made to pretend that it does not exist, as though it were abnormal or discreditable. But it serves no useful purpose to conceal it; rather the reverse. For if planning and administration are to be realistic, they must take departmentalism and its probable manifestations into account as part of the scene. Indeed, as I suggest, it should often be possible through intelligent anticipation to reduce those of its effects which are harmful.

This study has many limitations. To state some will not remove them, but will at least serve to warn the reader. In the first place, I have the biases of a convinced 'developer', believing that rapid economic development in the third world is desirable. I generally see problems from the point of view of the 'developers' rather than the 'developed', and would argue that relatively too much attention has been paid to social factors and too little to administrative factors in analysing the processes of development.

A second limitation is the sources of this study. On the positive side, I was able to visit settlement schemes in Ghana, Kenya, the Sudan, Tanzania and Uganda; to interview many people involved in settlement scheme organizations, and to administer a questionnaire to 100 junior staff on Mwea; to study primary documents in Ghana and Kenya, and secondary sources for eight countries—Ghana, Kenya, Nigeria, Rhodesia, the Sudan, Tanzania, Uganda and Zambia; and to hold long discussions with other students of settlement schemes. But against this I must record that many of my visits were excessively brief and the information gained superficial; that in applying the questionnaire to junior staff on Mwea a number of questions had to be abandoned (the shortcomings lying at least as much in the questions as in the responses); that primary documents usually have the disadvantage of omitting reference to the informal communication which is often crucial in determining the pattern of events; and that secondary sources are difficult to handle comparatively because the information they present is so often disparate.

Third, there is a major and very obvious methodological weakness. The approach has been to make an historical and administrative study of one scheme, the Mwea Irrigation Settlement, and then to follow it with comparisons with other settlement schemes, leading through to an attempt to draw together a comparative analysis and some practical lessons. Mwea has been used, as it were, for dissection in order to compile an inventory of the parts of a settlement scheme, and this inventory has then been used to provide boxes for sorting information about other settlement schemes. This approach has the obvious disadvantage of the implicit assumption that Mwea is typical of settlement schemes in general, and carries with it the danger of treating Mwea as a paradigm from which all other schemes are in some way lapses or deviations. Being aware of this shortcoming does not necessarily mean that it is sufficiently compensated for, and it is open to question whether, had another scheme been chosen for detailed study, somewhat different conclusions might not have emerged.

A fourth problem arises from the choices of research students. The schemes which get studied are usually schemes of studiable size. These tend also to be manager-sized, that is to say, of a size and complexity suitable for management by one man and a small staff. The very large schemes like Gezira and the Kenya Million-Acre Scheme are not as attractive, and tend to receive generalized studies which are not comparable in the level of detail with studies of smaller settlements. There is, therefore, a danger of an ideal type of the relatively small scheme emerging, even unconsciously, while larger schemes are ignored.

A final difficulty is the personality variable. A promising scheme may fail or an unpromising scheme succeed as a result of the performance of the man in charge. Further, personal idiosyncracies are often given free rein in the autonomous commands provided by many isolated settlement schemes. However, I have not described the personalities of individuals in any detail. This is partly because much of my perception is secondhand, through the eyes of others; partly because much of my understanding of people and events derives from interviews, and it would be contrary to the spirit of free discussion in those interviews to use them to give character sketches of third persons; and partly because I would not be sure that I was being fair. The resulting descriptive weakness is perhaps less serious than it might at first sight appear, for two reasons. In the first place, in general terms, I am able to discuss the character and attitudes of staff in chapters 7 and 8, and so to introduce this important variable into the analysis. Second, the attitudes, behaviour and conflicts which are described are to a considerable extent intelligible in terms of roles generated or demanded by evolving scheme situations. All the same, it has to be admitted, indeed emphasized, that within the bounds set by these roles and situations there is scope for individual variation and that at scheme level one of the most decisive influences remains the personality of a manager.

Acknowledgements

This study could not have been carried out without the help of many organizations and people. For reasons of space I cannot mention them all here, but I should like to thank them all none the less.

The Ministry of Overseas Development supported me during the bulk of the field work, and the Department of Government of the University of Manchester during most of the writing-up. I am grateful to both organizations, and to those in them who tolerated my absences while working on this study and who did the work that I might otherwise have done. I particularly want to thank the first and second Directors of the East African Staff College, Guy Hunter and Hamish Millar-Craig, for the understanding they showed towards my frequent disappearances, without which the field work could never have been completed.

Most of the research was carried out in Kenya. The President's Office approved my work, and I was given considerable assistance both in Nairobi and in the field by officials of several ministries and departments, especially the Provincial Administration, the Ministry of Agriculture, the Water Development Department, the Department of Settlement, and latterly, after its formation, the National Irrigation Board. In Kenya and England many of the principal participants in the early days of the Mwea Scheme were helpful in combing their memories and gave freely of their impressions and recollections. Among those officials who were serving at the time of this study, I owe particular thanks to Richard Alai who was District Commissioner for Kirinyaga District during much of the period, Mrs Luckham who was Librarian in the Ministry of Agriculture, and John Stemp, the Manager of the Perkerra Irrigation Scheme. On the Mwea Irrigation Settlement a great deal of assistance in many large and small ways was given by successive Managers, many members of the senior and junior staff, and tenants. Many of these submitted to long interviews and gave freely of their spare time to help. My greatest debt here is to E. G. Giglioli, now General Manager of the National Irrigation Board, for long and fruitful discussions and for consistent interest in and support for the work I was doing.

The opportunity to study the Volta River Resettlement Operation in Ghana arose from the initiative of Commander Sir Robert Jackson, and an invitation from the Ghana Government. While the purpose of my visit to Ghana in 1965 was a separate work on the Resettlements, the insight obtained into a West African settlement scheme provided a useful comparison with East Africa. I am grateful to E. A. K. Kalitsi, the Resettlement Officer, and G. W. Amarteifio, the Resettlement Manager, for long interviews and for the arrangements made for me to visit most of the resettlement towns; to James Moxon for his advice and hospitality; and to David Butcher, Laszlo Huszar and Rowena Lawson for discussions and encouragement. My especial thanks are due to Kwame Frempah, who as guide, assistant and friend, made my visit not only useful but enjoyable.

I am also obliged to those, not least the Managers, who enabled me to visit some Group Farms and the Nyakashaka Resettlement Scheme in Uganda, and the Upper Kitete Village Settlement in Tanzania.

This study has been made possible by the advice, insights, and findings of others. New lines of enquiry which I should otherwise have missed were opened up for me by Professor Colin Leys and Professor Gilbert White. A great deal of information and many ideas have come from other students of settlement schemes in East Africa. I find I cannot acknowledge in detail my many debts to them, but I have learnt much from discussions with Professor Raymond Apthorpe, Brack Brown, Simon Charsley, David Feldman, Caroline Hutton, Jon Moris, Peter Rigby, Stephen Sandford and Rachel Yeld, among others. To the Syracuse University Team working on village settlement in Tanzania—Kenneth Baer, Carol Fisher, Nikos Georgulas, James Lewton-Brain, Robert Myers, John Nellis, Anthony Rweyemamu, Garry Thomas and Rodger Yeager—I am grateful for stimulating discussions and for many personal communications. Those with whom I have corresponded are too numerous to list, but I hope they too will accept my thanks for information freely given. The sense of excitement in a common interest among those carrying out research on settlement schemes, and their generosity in making findings available, have maintained a sense of momentum and added greatly to the pleasure of investigation.

In the context of the Mwea Irrigation Settlement I am grateful to Stephen Sandford for information from his analysis of early expenditure on the Scheme; Benjamin Kihara for carrying out some of the interviews with junior staff; and James Kimani for obtaining detailed information which filled in some of the gaps in my knowledge. I owe most, however, to Jon Moris for his partnership in studying the Scheme. The UNICEF/Makerere Farm Innovation Survey carried out under his direction by students of the Department of Agriculture,

Makerere University College, has provided useful understanding. But my greatest debts to him are for long discussions to which he brought the insight of a sociologist widely familiar with agricultural development in East Africa, and in which many of the ideas in this book arose.

The experience of writing has taught me to appreciate the value of comments on drafts, and the amount of time and energy that making them entails. R. F. Marshall, Jon Moris, Wyn Reilly, Stephen Sandford, Douglas Taylor and William Tordoff read and commented on substantial sections, and Brack Brown, Cherry Gertzel, E. G. Giglioli, Professor A. H. Hanson and Professor W. J. M. Mackenzie read and commented in detail on entire drafts. Annuschka Reilly eliminated much bad style from an earlier version. Much reorganization and correction has resulted from the suggestions made by these various people to whom I am grateful for the time and effort they so generously devoted. Responsibility for errors of fact, interpretation and style that remain is, of course, entirely mine.

My greatest debt is to Professor W. J. M. Mackenzie for his support and encouragement from the beginning, for piloting through the complicated arrangements for the field work, for his detailed comments and wise guidance, and for making the writing-up possible. Above all, I am grateful to him for the stimulation of a range of ideas which made thinking and writing both more exacting and more rewarding.

Arline Harris and Tamie Swett brought imagination, stamina and forebearance to their battles with the typing. Rachel Higham managed to draw the first versions of the maps despite the attentions of her children, and A. G. Kelly drew the final versions.

Finally, I want to thank all those who helped me to remain tolerably sane for most of the time I was writing.

Part I
The Setting and the Schemes

SOME SETTLEMENT PROGRAMMES AND SCHEMES
IN EIGHT AFRICAN COUNTRIES

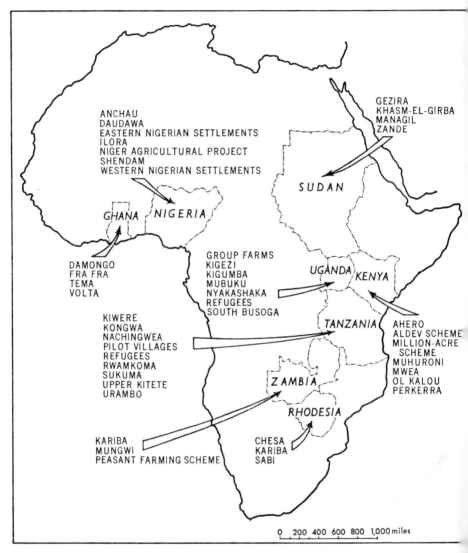

Note: Most of the titles of programmes and schemes have been abbreviated. The full titles are given on pages xv—xvi.

CHAPTER 1

The Significance and Study of Settlement Schemes

Scale and importance of settlement schemes

In the past decade, the balance of economic prescriptions for the third world has swung away from industrialization and towards agriculture as a means of achieving economic growth.[1] At the same time governments in less developed countries, whether still under colonial rule or independent, have launched projects for the reorganization of agriculture and for the introduction into agriculture of more advanced technology. Agricultural schemes started in British colonies before independence, and since independence in nations formerly under British rule, have included a wide range of approaches: land consolidation, farm planning, crop rotation, credit schemes, marketing co-operatives, cash crop introduction, mechanization and state farming, among others.

Some of the most conspicuous of these projects have been what are described as settlement or resettlement schemes. In the colonial period these were initiated in places as widely dispersed as Aden, Ceylon, Malaya, North Borneo, Sarawak and Trinidad.[2] In anglophone Africa (which is considered here notably in Western, Central and Eastern Africa including the Sudan) settlement and resettlement schemes have been and continue to be common and prominent.

The resources devoted to these projects have been large. They have, of course, included land, labour, and technical and administrative expertise; but an indication of the magnitude of investment is most readily given in terms of capital. Table 1.1 presents actual or planned expenditure for some of the largest settlement and resettlement programmes and schemes:

---

[1] See for instance, William H. Nicholls, 'The Place of Agriculture in Economic Development', in Carl K. Eicher and Lawrence W. Witt, eds., *Agriculture in Economic Development*, McGraw-Hill, New York, 1964, pp. 11–44.

[2] *Notes on Some Agricultural Development Schemes in British Colonial Territories*, H.M.S.O., London, October 1955.

TABLE 1.1

Orders of magnitude of capital costs of some settlement and resettlement schemes in anglophone Africa

Figures to the nearest million

Country	Scheme	Capital cost £ million
Ghana	Volta River Resettlements[1]	9
Kenya	Million-Acre Settlement Scheme[2]	23
	Mwea Irrigation Settlement[3]	1
Nigeria	Western Nigeria Farm Settlement Programme 1962–68[4]	6*
	Eastern Nigeria Resettlements 1962–68[5]	6*
Northern Rhodesia	Kariba Resettlement[6]	1
Southern Rhodesia	Resettlements under the Land Apportionment Act[7]	3
	Kariba Resettlement[8]	1
Sudan	The Gezira Scheme[9]	20
	Managil South-Western Extension of the Gezira Scheme[10]	S39
	Khasm-el-Girba Resettlement[11]	S25*
Tanzania	Village Settlements[12]	9*

*Forward estimates, not actual expenditures.

Notes: Accounting systems vary, and it is common for settlement schemes to receive substantial hidden subsidies which do not appear in official figures such as those quoted here.

Here and subsequently, wherever conversions into £ sterling have been made from other currencies, the exchange rates before the U.K. devaluation of 1967 apply.

[1] *Sources: Volta River Authority Annual Report for the Year 1965*, Guinea Press, Accra, 1966, p. 19. The final capital cost will be considerably larger.

[2] Communication, 10 April 1968, from the Acting Financial Controller (Settlement), Department of Settlement, Kenya, giving the capital cost to 30 June 1967 as £22½ million, and the estimated final capital cost for the extension to 1½ million acres as £28½ million.

[3] 'National Irrigation Schemes: Consolidated Capital, Expenditure, and Revenue Figures for the 11 Years to 30 June 1966 (amended)', (mimeo), by the Secretary/Chief Accountant, National Irrigation Board. The capital expenditure on the Mwea Irrigation Settlement for the period to 30 June 1966 was given as £787,368.

[4] *Western Nigeria Development Plan 1962–68*, Sessional Paper No. 8 of 1962 of the Western Nigeria Legislature, p. 68.

[5] *Eastern Nigeria Development Plan 1962–68*, p. 164.

[6] Montague Yudelman, *Africans on the Land*, Harvard University Press, Cambridge, Mass., 1964, p. 271.

[7] Elizabeth Colson, *Social Organization of the Gwembe Tonga*, Manchester University Press, 1960 (Appendix C: Kariba Resettlement—Northern Rhodesia: a note on Northern Rhodesian Settlement Plans by the Development Officer of the Northern Rhodesian Government), p. 221.

In 1967 and early 1968, in addition to these programmes and projects, many more were being considered or launched. In Ghana, for example, it was reported in 1967 that a settlement farm mainly for school-leavers was being started at Kpandu, with an estimated capital cost of £1·2 million.[1] In Western Nigeria the Crash Development Programme for 1968–1971 included an estimate of £3·2 million for farm settlements.[2] In Northern Nigeria a resettlement operation was in hand for the population displaced by the Kainji dam. In Kenya the Million-Acre Scheme was being extended to one and a half million acres. In Zanzibar a proposal was reported in 1967 to move 100,000 people, a third of the population, from less fertile to more fertile land.[3] Moreover, in several countries in Africa a good deal of *ad hoc* resettlement, with varying degrees of assistance from governments, from the United Nations High Commission for Refugees, and from voluntary agencies (among which Oxfam was prominent), continued to take place to deal with the severe refugee problems of the continent. In May 1967 there were estimated to be 740,000 refugees in Africa south of the Sahara, of whom 100,000 had been re-established as a result of rural implantation programmes, but of whom some 250,000 were still in need of assistance.[4] Further, in Nigeria it seemed likely that refugee and resettlement problems resulting from the civil war would continue for some time.

[1] *Ghanaian Times* (Ghana), 18 September 1967. The estimate given was 3·5 million New Cedis.

[2] *Crash Development Programme 1968–71*, Ministry of Agriculture and Natural Resources, Ibadan, 1968.

[3] *The Standard* (Tanzania), 23 March 1967, reporting a statement by the First Vice-President, Mr Karume.

[4] Reported in *Africa Research Bulletin, Political, Social and Cultural Series*, 1–31 May 1967, p. 775.

---

[8] Colson, op. cit. (Appendix D: Kariba Resettlement—Southern Rhodesia, by the Chief Information Officer, Native Affairs Department, Southern Rhodesia), p. 226.

[9] *Report of the East African Royal Commission 1953/55*, Cmd. 9475, London, H.M.S.O., 1955, p. 330.

[10] D. J. Shaw, 'The Managil South-Western Extension: an Extension to the Gezira Scheme', *Bulletin No. 9 issued by the International Institute for Land Reclamation and Improvement*, H. Veeman and Zonen N.V., Wageningen, 1965, p. 20.

[11] D. S. Thornton and R. F. Wynn, 'An Economic Assessment of the Khasm-el-Girba Project' (mimeo), Faculty of Agriculture, University of Khartoum, March 1965, p. 2.

[12] *Tanganyika Five Year Plan for Economic and Social Development 1 July 1964– 30 June 1969*, Dar es Salaam, Government Printer, 1964, Vol. 1, p. 21. Over sixty village settlements were to be established in the five year period, each justifying the investment in economic and social overheads of £150,000. In April 1966, however, it was announced that the programme had been suspended.

More imponderable, but potentially much larger in scope, are the long-term possibilities for settlement schemes, especially for irrigation. In Ghana a succession of surveys have been carried out in the lower Volta flood plain where there are some 300,000 acres of riverain land potentially suitable for irrigation, and various proposals for development have been, and continue to be, considered.[1] In Northern Nigeria a pre-irrigation soil survey of the extensive area of the Chad basin between Lake Chad and Kano is in hand.[2] In Uganda the Mubuku Irrigation and Demonstration Project, started in 1962, continues with a view to establishing the possibility of developing 15,000 acres for irrigation.[3] In Kenya the Ahero Pilot Irrigation Project of 2,000 acres is being pushed forward with outstanding energy and speed in an area, the Kano Plains of Nyanza Province, where the ultimate irrigation potential is about 35,000 acres.[4] Also in Kenya, investigations continue in the lower Tana Basin where a United Nations survey has reported that an irrigation scheme of 360,000 acres might cost £150 million.[5] In Tanzania a United Nations team has carried out an irrigation survey of the Wami and Pangani valleys, and the Rufiji valley has been estimated to have an irrigation potential of 200,000 acres.[6] In Malawi irrigation investigations are under way in the Shire Valley. In Botswana funds were provided in 1967 for a survey of the potential of the huge area of the Okovango swamp.[7] Elsewhere many investigations and new projects are being carried out: in the Blatte Valley in Ethiopia, in the Kafue basin in Zambia, in the Sabi valley in Rhodesia, at Vuvulane in Swaziland and Qamata in the Transkei in South Africa, to mention but a few. Moreover, there are many similar surveys and schemes in the francophone African countries; and in North Africa what is perhaps the largest project under consideration in Africa, a proposal to irrigate an ultimate 740,000 acres in the Sebou Valley in Morocco, has been estimated to cost £210 million.[8] The list could be extended, but the point is already clear: settlement and resettlement schemes

[1] See, for instance, *West Africa*, No. 2586, 24 December 1966, p. 1475, and *Daily Graphic* (Ghana) 18 and 19 August 1967.
[2] *New Nigerian* (Kaduna), 1 August 1967.
[3] G. D. Agrawal, 'Some Considerations Affecting the Organization of the Mubuku Irrigation Scheme' (mimeo), Makerere University College, Rural Development Research Project paper no. 35, 1966.
[4] E. G. Giglioli 'Aspects of Planning Irrigated Settlement in Kenya' (mimeo), Nairobi, National Irrigation Board, 28 August 1967.
[5] Interim report of a United Nations Special Fund survey of the lower Tana basin in Kenya, reported in the *East African Standard* (Nairobi), 17 November 1965
[6] Report of a United States Bureau of Reclamation study team, quoted in the *Daily Nation* (Nairobi), 1 August 1967.
[7] Broadcast in English from Gaberones, Botswana, 26 July 1967.
[8] *Financial Times* (London), 9 March 1967.

in Africa have involved large investments in the past, and the scale of future investment is potentially enormous. In social and economic terms, however, the record of past schemes has been discouraging. Not only have they given rise to many problems, but outright failures and collapse have been common. In the Gold Coast, for instance, the Gonja Development Company, formed in 1950 with a share capital of £1 million and with village settlement as one of its aims, was wound up in 1956 having achieved, in the words of one authority, 'nothing except extravagant expense for the Ghana Government'.[1] In Nigeria the Niger Agricultural Project at Mokwa was closed down in 1954 after an expenditure of £½ million.[2] In the southern Sudan the Zande Scheme, in which there had been a government investment of £1 million, culminated in 1955 in riots and martial law.[3] In Uganda the South Busoga Resettlement Scheme more or less ceased to exist in 1961,[4] and in Kenya the Lambwe Valley Settlement Scheme, started in 1951, was a virtual failure by 1959.[5] Nor has this disappointing experience been limited to the colonial period. In Tanzania, the ambitious post-independence programme for village settlements quickly ran into difficulties and was suspended in 1966,[6] and in Western Nigeria the settlements for school-leavers experienced multiple problems. Indeed in almost all countries settlement schemes have been criticized by sociologists, agriculturalists and economists for failure to achieve their social, agricultural and economic objectives, and for their absorption of scarce resources which might have been put to better use. In view of the common failures of the past and the extensive proposals for the future, there is a case for trying to learn from the experience to date and for examining the nature of settlement schemes in order to show what is being decided upon, so that any alternative policies can be more realistically weighed.

[1] S. La Anyane, *Ghana Agriculture, its Economic Development from Early Times to the Middle of the Twentieth Century*, Oxford University Press, Accra, 1963, p. 171.

[2] K. D. S. Baldwin, *The Niger Agricultural Project, an Experiment in African Development*, Blackwell, Oxford, 1957, p. 183.

[3] Conrad Reining, *The Zande Scheme: an Anthropological Case Study of Economic Development in Africa*, Northwestern University Press, Evanston, Illinois, 1966, pp. 214–217.

[4] Susan J. Watts, 'The South Busoga Resettlement Scheme', *Syracuse University Program of Eastern African Studies Occasional Paper No. 17*, Maxwell Graduate School of Citizenship and Public Affairs, Syracuse, April 1966, p. 36.

[5] *African Land Development in Kenya, 1946–1962*, Ministry of Agriculture, Animal Husbandry and Water Resources, Nairobi, 1962, pp. 178–181.

[6] Press release, IT/I.302, Ministry of Information and Tourism, Dar es Salaam: Address by the Second Vice-President, Mr R. M. Kawawa, at the opening of the Rural Development Planning Seminar at the University College, Dar es Salaam, 4 April 1966.

Much of the experience gained in settlement schemes and similar projects was sifted and digested in the 1950s when a body of doctrine about their successes and failures began to become established. In a perceptive article in 1954 Arthur Lewis emphasized the importance of a number of factors, many of which affected the performance and motivation of settlers.[1] In 1957 Baldwin exposed the sociological, agricultural, and economic shortcomings of the Niger Agricultural Project.[2] In 1959 Gaitskell contributed his classic work on Gezira with its wise precepts about development.[3] It became recognized that the post-war schemes had often been over-ambitous; that experiment and gradualism were important; that mechanization was difficult in tropical conditions and in underdeveloped countries; and that peasant settlers were more rational in their behaviour than had often been supposed. Something of an orthodoxy of explanation and understanding became accepted.

Strangely little attention was paid, however, to the creation, organization and management of settlement schemes.[4] Explanation of failure or success was generally limited to the dimensions of land, climate, settlers and agricultural process. Staff and organizations were taken as given. Only in Reining's admirable study of the Zande Scheme,[5] published in 1966, have they been examined in any detail by someone who was not himself involved in management. Reining's suggestions that 'the technological officials and experts be as much part of the research as the people being developed', and that 'the totality of the development project should be studied, rather than just the people being developed',[6] represented a new and useful approach.

It was only in pursuing his search for an understanding of what happened to those who were being developed that Reining came to study the developers. But *a priori* it would appear that in trying to understand projects and to derive practical lessons from them the staff and their organizations are, if anything, more important than the people they affect. It is the staff who decide policy and execute it. It is they who perceive or fail to perceive the details of the situation

[1] W. Arthur Lewis, 'Thoughts on Land Settlement', *Journal of Agricultural Economics*, Vol. 11, June 1954, pp. 3–11, reprinted in Eicher and Witt, op cit., pp. 299–310.
[2] Baldwin, op. cit.
[3] Arthur Gaitskell, *Gezira, a Story of Development in the Sudan*, Faber and Faber, 1959.
[4] Gaitskell's treatment of Gezira is an exception. Gezira was, however, in some respects a special case (see pp. 18–19) and Gaitskell had the disadvantages, as well as advantages, of writing from personal managerial experience.
[5] Reining, op. cit.
[6] Ibid., p. 231.

in which a scheme is launched. It is they, and not the people being developed, who hold the initiative, especially in the early stages of a project. If staff and organizations are examined, not as they were by Reining as a secondary concern but as a primary focus, then more practical lessons may emerge. Developers may be able to learn something about their own behaviour, about the problems and conflicts they are likely to face and about needs that have to be anticipated.

But if the developers were to become the exclusive object of any study, an unbalanced and unrealistic view would emerge. If valid lessons are to be teased out, then settlement schemes must be examined as wholes, including not only the developers and the developed but also land, climate, infrastructure, economic processes, and the social, political and economic environments in which they are found. Moreover, changes over time must be taken into account. Consideration of these aspects is attempted here, however inadequately, within the loose limitations of analysing settlement schemes primarily as organizations. This approach has several practical advantages. It directs attention to those administrative aspects which have tended to be ignored. It implies examination of the systematic relationships of parts, so that it can include all the elements, relationships, and activities which are found to be relevant. The roles, attitudes and behaviour of both staff and settlers are thus included, much as would be those of management and workers in a study of a factory. Equally, the physical layout and economic processes of a settlement scheme can be taken into account much as they would be in a study of an industrial organization.

Many questions are suggested by this approach. What sorts of organizations create settlement schemes, and what are their relationships with one another? What sorts of men become staff and settlers, and what attitudes and expectations do they bring to schemes? How do they behave towards each other, and what adaptations do they make? What forms of organizations develop on settlement schemes and why? What relationships do they have with other organizations in their environment? How do schemes and their problems change over time? What types of scheme can be identified? What lessons can be learnt and applied to future settlement schemes and to other development projects? What criteria does this approach suggest should be applied in evaluating agricultural schemes in general? These are some of the questions which are central to this study. But before considering them, there are problems of definition and method which, while of less interest to a practical man than to a student, must be dealt with in order to set more clearly the scope and direction of what follows.

9

Scope and some limitations of the study

Up to this point settlement and resettlement schemes have been treated as though they are an intelligible and coherent field of study. But are they? And if so, how should the field be defined?

On first inspection, the entities described as settlement or re-settlement schemes display a bewildering variety. Among their purposes they may include soil conservation, the relief of famine, tsetse fly clearance, the settlement of refugees, evacuees, unemployed and school-leavers, the creation of new production, ideological aims, and the solution of political problems. In size they range from the Gezira and Managil Schemes in the Sudan, which together cover some 1,800,000 acres[1] and have supported a population of nearly 900,000,[2] to the Nyarushanje Farm School Settlement in Uganda, which in 1965 covered 160 acres and supported six unmarried young men.[3] In capital cost they have varied from £4,000 per settler as planned for the Western Nigeria Farm Settlement Programme[4] to just over £1 per head of population claimed for the Kigezi Resettlement Scheme in Uganda.[5] The land settled has sometimes been uncultivated bush or forest, and in other cases was already cleared and developed before settlement. The agricultural systems have ranged from family smallholdings farmed traditionally to highly centralized and mechanized cash crop farming. In some schemes almost all operations have been left to the self-help and initiative of the settlers; in others many operations have been carried out by a settlement organization. The diversity is such that it may well be asked whether settlement schemes constitute a class of entity that hangs together as an intelligible field of study, or whether the use of the expression 'settlement scheme' to describe them all leads to the old error of supposing that because there is a term there must be a natural category to which the term refers.

Three definitions of settlement or resettlement schemes illustrate this difficulty. Bridger has used 'settlement schemes' to mean 'the transfer of population from one area to another on a planned basis,

[1] R. J. Harrison Church, 'Observations on Large Scale Irrigation Development in Africa', *Agricultural Economics Bulletin for Africa*, No. 4, November 1963, p. 35.

[2] Shaw, op. cit., p. 16. The exact figure, 887,329, was derived from First Population Census of Sudan, 1955/56, Khartoum, Ministry of Social Affairs, Population Census Office, 1957, Table 7.

[3] Personal visit, January 1965.

[4] Dr D. S. Onabamiro, Minister of Agriculture and Natural Resources, reported in *Western Nigeria House of Assembly Debates Official Report*, 16 April 1964, Column 546.

[5] J. W. Purseglove, 'Kigezi Resettlement', *The Uganda Journal*, Vol. 14, No. 2, September 1950, p. 149.

the object being to raise living standards'.[1] Belshaw has written that 'All "resettlement schemes" may be defined as projects involving the planned and controlled transfer of population from one area to another'.[2] Apthorpe has suggested, with careful qualifications, that land settlement can be regarded as one subdivision of 'planned social change that necessarily does entail population movement, population selection and most probably population control subsquently', there being at least two other subdivisions—the sedentarization of nomadic or pastoral populations, and the villagization of cultivators.[3] In interpreting these three definitions there are varying degrees of difficulty in determining what mixture they represent of rationalizations of common usage, logical definitions, descriptions of natural categories, or convenient *ad hoc* outlines of the scope of the papers they introduce.

The purpose here is neither to describe a natural category nor to present an exclusive verbal definition, but simply to outline the boundaries of the inquiry. For convenience the main area of attention can be defined by common usage. On examination, what are called settlement or resettlement schemes do in fact show the two features common to the three definitions or descriptions just quoted: a movement of population; and an element of planning and control. These correspond roughly with the words 'settlement' (or 'resettlement'), and 'scheme', respectively. Both 'settlement' and 'resettlement' imply population movement, but since 'settlement' is the more inclusive word, it will be used here except where it appears in the title of a scheme or where the particular sense of dislodgement before transfer implied in 'resettlement' is required. But 'settlement' means more than this. Both as a word and in the schemes considered, there is an implication of the establishment of people upon land in some relatively permanent manner. The second word, 'scheme', is synonymous with 'project', and is commonly used to imply an organized attempt to introduce change. In anglophone Africa schemes have been initiated by agencies, usually but not always governmental, which can be regarded as distinct from the receiving environments, and have typically involved some degree of continuing influence and control, if only for a limited period. What we find described as settlement schemes corresponds with the overlap

[1] G. A. Bridger, 'Planning Land Settlement Schemes', *Agricultural Economics Bulletin for Africa*, No. 1, September 1962, p. 21, footnote.
[2] D. G. R. Belshaw, 'An Outline of Resettlement Policy in Uganda, 1945–63', *East African Institute of Social Research Conference Paper June 1963*, p. 1.
[3] Raymond Apthorpe, 'A Survey of Land Settlement Schemes and Rural Development in East Africa', *East African Institute of Social Research Conference Papers January 1966*, No. 352, pp. 1–2. The quotation is from page 1.

of the words 'settlement' and 'scheme' in these senses. They do, in practice, involve both population movement and organized attempts to establish people upon land.

This outline of meaning does not assert that settlement schemes are a natural category; it is, rather, a statement of focus. It is not meant to exclude from examination related phenomena, especially schemes such as the Groundnut Scheme, which do not involve settlement, and settlement, such as population movements from higher to lower rainfall areas, which do not involve anything that can be called scheme. This broad span of relevance is justified by the primitive state of theory about settlement schemes in particular and about the organization of agricultural development in general. The common forms of description of agricultural projects and programmes often cut across settlement schemes. Land reform, in Warriner's sense of 'the redistribution of property in land for the benefit of small farmers and agricultural workers',[1] would include the Kenya Million-Acre Scheme but not, for instance, the Kiwere Settlement in Tanzania where unoccupied land was colonized. Similarly, descriptions of agricultural schemes by crop or by production technique may include both settlement and non-settlement. For instance, mechanical agriculture includes tractor hire services, state farms, most large estates, and some but not all settlement schemes. Another difficulty emerges over deciding whether borderline cases are settlement schemes or not. For instance, land consolidation in the Central Province in Kenya is not normally regarded as a settlement scheme, yet it involved 'scheme' in the reorganization and re-allocation of holdings, and 'settlement' when people were encouraged to move out of villages on to the land. In view of these various problems of definition, it is more useful to be pragmatic and inclusive than doctrinaire and exclusive in deciding what is relevant.

Much of the evidence used is the work of others who have studied settlement schemes. Inevitably the interests of different disciplines mean that information from one scheme is not often of the same sort as information from another. It can be said of settlement schemes, as Warriner has of land reform, that they constitute:

> an academic no man's land. No single science or study has yet established its claims, and each has its limitations. No single method of approach can take us all the way. The subject remains what the Americans call inter-disciplinary and the English call borderline.[2]

[1] Doreen Warriner, 'Land Reform and Economic Development', in Eicher and Witt, op. cit., p. 272.
[2] Ibid., p. 274.

Fortunately for this study, the academic no man's land of settlement schemes has been attractive to many people by virtue of its visibility, its relatively clear boundaries, and its subdivision into individual schemes which make convenient units for study. Indeed, this attraction has made students some of the later invaders that have entered the territory colonized by settlement schemes. As a result the literature is rich and growing. Despite this attention, however, it cannot be said that there is any organized body of theory about settlement schemes.[1] Each student has brought his own discipline's point of view to bear, and has left some areas unilluminated. Just as colonizers of unoccupied land have to provide for themselves facilities which they would elsewhere take for granted, so students of settlement schemes have sometimes had to spread their inquiries beyond the normal limits of their disciplines. Where this has been done the results have often been particularly useful. For example, Baldwin, through discussing sociological as well as agricultural and economic aspects of the Niger Agricultural Project, has contributed many revealing insights.[2] Roder, though a geographer, by considering historical and administrative factors has made his account of the Sabi Valley Irrigation Projects far more intelligible than it would have been had he limited himself to more conventionally geographical description.[3] Gaitskell's masterly survey of the history of Gezira is similarly valuable because of its breadth.[4] In addition, many shorter works by social anthropologists, sociologists and economists have also shot shafts of light into the subject, and from different angles.

The result is an exciting but unmanageably large mass of disparate evidence. When in 1966 a number of students of settlement schemes in East Africa tried to draw up a checklist of questions that might be asked about settlement schemes in trying to see them as intelligible wholes,[5] it became clear that it would be impossible for any one scheme to discover all that was relevant. Only by being selective is it possible to proceed at all. This inevitably entails high risks of error and distortion. This study is selective, and unbalanced, in regarding

---

[1] To the best of my knowledge at the time of writing, April 1968.

[2] Baldwin, op. cit.

[3] Wolf Roder, *The Sabi Valley Irrigation Projects* (University of Chicago Department of Geography Research Paper No. 99), University of Chicago Press, Chicago, 1965.

[4] Gaitskell, op. cit.

[5] 'A Tentative Checklist of Questions about Settlement Schemes' (mimeo), ('based on initial discussions of Brain, Charsley, Chambers, Robertson and Yeld at Makerere on 5 January 1966, and subsequent discussions of Apthorpe, Brown, Chambers, Moris, Myers, Nellis, Rigby and Yeld, with assistance from Etherington, at Nairobi on 4 and 5 February 1966. Edited by Hutton and Apthorpe, following pre-editing by Chambers.')

settlement schemes primarily as organizations, and the reader must weigh for himself the degree of distortion this involves.

Two further limitations have been placed on the scope of this study. In the first place only Ghana, Kenya, Nigeria, Rhodesia, the Sudan, Tanzania, Uganda and Zambia are considered, since these are the only countries in colonial and ex-colonial anglophone Africa in which settlement schemes appear to have been examined in any detail. Second, European settlement is not considered, since it would have introduced extra variables into a study which was already in danger of trying to handle too wide a range of phenomena.

The sequence of presentation is first to outline the background and origins of settlement schemes in these eight countries, including brief mention of most of those which are reasonably well documented. Then in Part II a particular scheme, the Mwea Irrigation Settlement in Kenya, is examined in detail as an administrative case history. In Part III settlement schemes are considered as organizations in which people—staff and settlers—are actors, in which the parts are systematically related, and which exist and survive through relationships with other organizations in their environments. Finally, in Part IV, an attempt is made to draw together the analysis in forms which lead to practical lessons, by describing changes which take place in settlement schemes over time, by suggesting some types of settlement scheme which can be separated out, and by outlining policy implications for the future.

CHAPTER 2

The Background and Origins of Settlement Schemes in Africa

Although settlement schemes in tropical Africa have shown great diversity and have arisen out of many different sets of circumstances, there are some elements and trends in their backgrounds and origins which are particularly common and significant. What may be general, and what particular, about them can be understood in terms of the conditions which have given rise to them. As elsewhere in the world, the most obvious and important of these conditions have been physical environment, people, and government, and the processes of change in which these have together been involved. This chapter outlines some of the more salient aspects of the relevant physical, human and governmental background of tropical anglophone Africa, and then describes the changing origins and purposes of settlement schemes in the colonial period and in independence. In doing this, mention will be made of most of the schemes which will be discussed later in Chapters 7 to 12.

Land, people and government

Although all the settlement schemes that will be considered lie between the tropics of Cancer and Capricorn, the physical environments in which they are located vary widely. At one extreme, the Gezira triangle south of Khartoum in the Sudan has a low and uncertain rainfall; at the other extreme, tropical rain forests are found in parts of central Ghana, southern Nigeria, southern Sudan and southern Uganda. Between these two extremes of climate and vegetation are the tropical savannahs of northern Ghana and northern Nigeria, parts of southern Sudan and northern Uganda, and much of Kenya, Tanzania, Zambia and Rhodesia. While there is a wide range of conditions within this savannah zone, it is typified by wet and dry seasons each year, by light, easily erodable soils, and by vegetation varying from thick woodland to grassland with scattered bush. Infestation by tsetse fly is also common. Although settlement schemes are found both in near-desert conditions and in tropical rain forests, it is in the tropical savannah that most of them have been initiated.

The evolution and form of settlement schemes have been influenced by many social factors. Among those which appear significant are traditional social and political organization, varying from the hierarchical, centralized systems found in West Africa and parts of Uganda and the Rhodesias, to the less centralized 'chiefly' systems of many parts of Tanganyika and Nyasaland, and the segmentary societies common in East Africa, particularly Kenya; patterns of residence, in West Africa commonly in nucleated villages, and in East Africa commonly on family homesteads; land tenure, in which individual ownership of land in the European sense scarcely existed as land was traditionally vested in the community; and the family as the basic economic unit.

Before the later stages of colonial rule most of the people were at subsistence or near-subsistence levels, with economies varying from nomadic pastoralism associated with dry conditions to settled agriculture associated with high rainfall. In the tropical savannah belt there were mixed pastoral and agricultural economies. Cultivation in these mixed economies was typically shifting, relying upon long bush fallows to restore the fertility of the soil between successive periods of cultivation.

The impact of British colonial rule was different in different places. Where there was a coherent indigenous authority, indirect rule through that authority was preferred to direct administration. Where there was no clearly organized indigenous authority an attempt was often made to create one. There was a contrast, too, between territories in which there was substantial European settlement, such as Kenya and the Rhodesias, and territories without such settlement, there being generally closer administration in the former. Nevertheless, there were considerable similarities in the structure and style of field administration. 'The Administration', as it was commonly known, consisted of a hierarchy of administrative officers, typically with a Provincial Commissioner (P.C.) in charge of a Province, a District Commissioner (D.C.) in a District, and a District Officer (D.O.) at a sub-district level.[1] It was usual for the administrative officer to have broad authority to maintain law and order, collect taxes, promote welfare, act as a magistrate, and deal with all government matters that could be described as political. Although the social and educational homogeneity of those who became administrative officers can easily be exaggerated, it is true that many had been to public schools and Oxbridge. In understanding their behaviour, however, it may be at least as important

[1] The Administration in Kenya is described in greater detail on pp. 49–52.

that certain types of person were attracted by the colonial situation,[1] and that similar situations of power and of social and cultural isolation induced in them similar paternal and autocratic attitudes.

In studies of British colonial government there has been a remarkable tendency to focus attention on administrative officers to the neglect of the officers of technical departments such as Agriculture, Works, Education and Health.[2] This may partly be explained by a romantic fascination with the D.C. in his Bombay bowler, bush jacket, shorts and long stockings, encouraged by such contributors as Edgar Wallace to the literature of colonialism. Further, administrative officers have been more articulate than their technical colleagues in writing their memoirs. Moreover, in the earlier, and to the European imagination often more engaging, stages of colonial rule, administrative officers were more powerful, more conspicuous, and relatively more numerous. But whatever the explanations of the neglect of technical officers, the plain fact is that in the later stages of colonial rule they considerably outnumbered the Administration in the field and progressively exerted more influence at all levels of government. Not surprisingly interdepartmental tensions became a common feature of government, but they have not been a subject of specific systematic study.[3] Indeed, they are cloaked, in the rather limited literature, by occasional references to splendid co-operation, which writers possibly find more worthy of remark because of its unusualness. Too much could be made of this, but in attempting to understand the origins and progress of settlement schemes, tensions and rivalries between administrative and technical officers, quite as much as co-operation, need to be taken into account.

[1] For discussion of the mentality of those who were attracted to the colonial situation see O. Mannoni, *Prospero and Caliban, the Psychology of Colonization*, Praeger, New York, 1956. Although Mannoni's assertion (p. 104) that 'no one becomes a real colonial who is not impelled by infantile complexes which were not properly resolved in adolescence' appears extreme, his suggestion that colonial situations appealed to people with misanthropic tendencies does carry some conviction.

[2] See Robert Heussler, *Yesterday's Rulers, the Making of the British Colonial Service*, Syracuse University Press, 1963; A. L. Adu, *The Civil Service in New African States*, George Allen and Unwin, 1965; and Richard Symonds, *The British and the Successors, a Study in the Development of the Government Services in the New States*, Faber and Faber, 1966. Heussler gives a somewhat misleading impression of the Colonial Service by almost completely ignoring the technical departments. While Adu and Symonds do mention them, they do not accord them as much prominence as the Administration.

[3] Among studies of settlement this aspect has, however, recently been recognized by Reining and Illingworth. See Reining, op. cit., pp. 102–103, 131–132, and 199–203, and Susan Illingworth, 'Resettlement Schemes in Uganda: a Case Study in Agricultural Development', unpublished M.A. thesis, Nottingham University, 1964, pp. 142–155.

17

It would be out of place here to describe in detail the impact of colonial rule. Those aspects which are most relevant will be touched on as the study proceeds. Suffice it to say that the effects of the imposition of a greater degree of peace, and the progressive introduction of medical and veterinary services, provided conditions for rapid increases of human and animal populations which disturbed previous ecological balances; that Christian missions, the introduction of alien law, education, the growth of a cash economy, labour migration and other influences weakened the hold of custom; and that, with the exception of Rhodesia, African national politics led to the achievement of legal independence before 1965. All this may be familiar enough, but it is useful to recognize these various background factors which directly and indirectly influenced the creation and form of settlement schemes.

Colonial settlement schemes

Early settlement operations were described as resettlement, and were large-scale moves of people who were living, both before and after the moves, at a subsistence or near-subsistence level. The purposes of these operations were either political or humanitarian. Movements of population for political reasons were organized in Southern Rhodesia and Kenya in order to establish European settlement, and more widely in order to separate tribes into different areas. Humanitarian moves were carried out to evacuate areas infected with sleeping sickness, and took place, for example, along the Uganda coast of Lake Victoria in 1906,[1] in land south of the Zambezi in 1913,[2] among the Azande in southern Sudan in the 1920s,[3] and in the Tabora area of Tanganyika in the late 1930s.[4] Official services and agricultural planning can hardly have been other than minimal during the earlier operations, which are perhaps more usefully regarded as induced migrations rather than organized settlement schemes.

The first major operation that can properly be described as a settlement scheme has also become the largest. The Gezira Scheme[5]

[1] Illingworth, op. cit., p. 11.
[2] Thayer Scudder, *The Ecology of the Gwembe Tonga*, Manchester University Press, 1962, p. 154.
[3] Reining, op. cit., p. 101.
[4] *Report and Proceedings of the Conference of Colonial Directors of Agriculture held at the Colonial Office, July 1938*, Colonial No. 156, H.M.S.O., London, 1938, p. 123.
[5] The history of the Gezira Scheme is outlined in Lord Hailey, *An African Survey, Revised 1956, a Study of Problems Arising in Africa South of the Sahara*, Oxford University Press, 1957, pp. 1,010–1,014. The most authoritative and detailed account is that by Gaitskell, op. cit.

in the Sudan overshadows all others. It was conceived and brought to birth by the Administration of the Sudan in 'a spirit of paternalism and a paramountcy of native interest'.[1] At the turn of the century the development of the Sudan posed immense problems. In the style of grand imperial endeavour, a proposal for irrigation of cotton with the waters of the Blue Nile sought to link the interests of Lancashire with those of the people of the Sudan. With a deliberation which makes many modern schemes appear irresponsibly rushed, plans worked out in the first decade of the century led to a pilot pump scheme in 1910 and the completion of the Sennar dam in 1925. A tripartite agreement was patiently negotiated between government, a commercial managing syndicate, and the landowners, and by 1962 some 1,800,000 acres of near-desert had been transformed into highly productive land and what has been described as 'Africa's most impressive man-made landscape'.[2] Gezira became so much the backbone of the economy of the Sudan that it has even been considered that without it there could have been no Sudanese independence.[3] Further, by demonstrating the possibilities in the circumstances of the Sudan of large-scale irrigation tied in with mechanization, the scheme encouraged observers elsewhere in Africa to launch out on ambitious projects. It has remained however, *sui generis*, an anomalous giant which has not yet been imitated successfully on a similar scale anywhere in tropical Africa.[4]

Before the Second World War, in territories other than the Sudan, large-scale settlement with a radical change of agricultural system was not attempted. In part this was related to lack of finance: the Colonial Development Fund formed under the Colonial Development Act of 1929 empowered the British Government to spend £1 million a year on colonial development,[5] a sum which did not provide much scope for high capital enterprises. At the same time agriculturalists were pre-occupied with the groundwork of training staff, produce inspection, agricultural education, field experiments and the

[1] Gaitskell, op. cit., p. 30.

[2] Harrison Church, op. cit., p. 35.

[3] George H. T. Kimble, *Tropical Africa*, The Twentieth Century Fund, New York, 1960, Vol. 1, p. 190.

[4] The next largest irrigated settlement scheme in tropical Africa is that of the Office du Niger in Mali where, although the irrigable potential had been reckoned as 2½ million acres, by 1964 the area under irrigation was only about 120,000 acres, and the capital invested amounted in 1960 to 44,000 million C.F.A. francs (rather more than £60 million). John C. de Wilde, *et al.*, *Experiences with Agricultural Development in Tropical Africa*, Vol. 2, The Johns Hopkins Press, Baltimore, 1967, pp. 245–300 for a description and analysis of the scheme.

[5] Barbu Niculescu, *Colonial Planning, a Comparative Study*, George Allen and Unwin, 1958, p. 59.

control of crops and diseases, rather than with introducing radical change into agricultural systems. However, even in the early 1930s a concern about stability in African agriculture (which lies behind many colonial settlement schemes) had already appeared in proposals for irrigation to combat famine and to provide regular food supplies. For example, the Sabi Valley Irrigation Projects in Southern Rhodesia, of which there were eventually ten, most of them started during the 1930s, were justified at first as a means of providing grain in years of food shortage as an alternative to famine relief.[1] Similarly, in Kenya, after the indigenous irrigation system of the Njemps had been destroyed by a flood in the Perkerra river and famine had followed, a succession of proposals was put up in the 1930s for an irrigation scheme, the main object of which would have been to remove the need for government to provide food for the local population in bad years. It is symptomatic of the shortages of resources, notably finance and staff, that the scheme was not started until 1953.

Other proposals for less ambitious settlement projects arose out of another concern with stability. The later 1930s and early 1940s were the heyday of alarm about soil conservation in Africa. It is a commonplace today that increasing population pressure led to overgrazing and overcultivation; as the bush fallows of shifting cultivation became shorter, they no longer adequately restored fertility to the soil. But this was only gradually realized. In the words of the Director of Agriculture in Tanganyika in 1937:

> Prior to the depression, agriculturalists in Tanganyika, at least, were but little concerned with soil erosion, as prices of most crops were so high that there was little or no incentive to efficiency in production. Yields of crops declined, yet still gave a satisfactory profit. It was not until the onset of the economic depression, with its severe fall in price level, that producers were forced to acknowledge the great loss in soil fertility that had occurred, and that to counteract the fall in prices for agricultural products it would be necessary not only to maintain the soil fertility but to improve it.[2]

An alarmist view of the situation was the easier to accept because of ignorance about soils and about methods of maintaining fertility. After the first hopes that tropical forest soils would prove extremely fertile had been destroyed, a pessimistic view prevailed: tropical soils were considered thin and inferior to those of temperate zones. But as Worthington has pointed out, there are some tropical soils which in a temperate climate would be too clayey to allow cultivation, but

[1] Roder, op. cit., pp. 105–108.
[2] E. Harrison, *Soil Erosion*, Crown Agents, London, 1937, p. 3.

which are friable in tropical conditions;[1] and the pendulum has swung back towards a median view of soil fertility in tropical Africa with the recognition that the abundance of energy from sunlight and the range of root systems and water storage can compensate for deficiencies in the soil and permit a more rapid growth of plant life than in temperate regions.[2] However, in the 1930s and 1940s the ignorance about soils was compounded in its effects by the patent instability in new conditions of traditional agricultural systems. Lord Hailey wrote in 1938 that the most urgent problem in all territories was the introduction of methods which would maintain soil fertility without recourse to shifting cultivation,[3] but for a long time it was not clear how far intensive agriculture could replace shifting cultivation without depleting the soil.[4]

Three factors in particular concentrated attention on soil erosion. First, there was the objective situation of spectacular sheet and gully erosion in many areas, especially in the tropical savannah. Second, there was a community of ideas and instruction between different territories. Common ideas and attitudes were developed and imparted through the training of colonial agricultural officers at the Imperial College of Tropical Agriculture in Trinidad; through the Devonshire courses conducted at Oxford and Cambridge for administrative officers; and through exchanges of personnel between colonies. In addition, a degree of metropolitan control and influence was exerted through despatches from the Secretary of State to the Colonial Governors and through the visits and advice of the Agricultural Adviser at the Colonial Office and his staff. Colonial Directors of Agriculture also held at least two conferences in the 1930s. Significantly neither soil conservation nor settlement were in the agenda or resolutions of the conference they held in 1931, but in the 1938 conference both were considered at length.[5] A third factor which directed energy and attention towards erosion is less precise but perhaps equally important. Conservation had an emotional appeal. Conservationists found it easy to muster passion for their cause and what became something of a gospel of soil conservation inspired

---

[1] E. B. Worthington, *Science in the Development of Africa, a Review of the Contribution of Physical and Biological Knowledge South of the Sahara*, C.S.A./ C.C.T.A., n.d., but from internal evidence probably 1958, p. 139.

[2] Ibid., pp. 138–140.

[3] Lord Hailey, *An African Survey, a Study of Problems Arising in Africa South of the Sahara*, Oxford University Press, 1938, p. 960.

[4] *Agriculture in the Colonies*, Colonial Office, London, July 1947, pp. 27–28.

[5] *Report and Proceedings of the Conference of Colonial Directors of Agriculture held at the Colonial Office, July 1931*, Colonial No. 67, H.M.S.O., London, 1931; and *Report and Proceedings of Colonial Directors of Agriculture held at the Colonial Office, July 1938*, op. cit.

some ardent apostles. The vocabulary of conservation, full of words like destruction, reclamation, reconditioning, rehabilitation and redemption, both expressed and evoked strong feelings about a visible and serious situation. But possibly some who used these expressions found a certain satisfaction that irresponsible Africans could only be saved from destroying their environment and their source of livelihood by the enlightened measures devised by the administrator or agricultural officer; and the fact that the implementation of these measures was usually compulsory, requiring disciplined organization and an access of additional power and authority to those who carried them out, probably had some connection with officials' enthusiasm for what often became a crusade.

Against this background it can be understood that although resources were scarce, the removal and resettlement of population from eroded areas was proposed as a solution. In Kenya in 1935 a circular invited details of projects for water supplies and included among the possible purposes opening up new land for tribal expansion.[1] In about 1938 Colin Maher, the Officer in charge of the Soil Conservation Service in Kenya, reported of one tribe, the Mbere, that if they were to make any economic progress they would have to be moved from the eroded land they were occupying and settled in a more favourable area.[2] But before the Second World War, although such ideas were in the air, little resettlement that was also scheme appears to have taken place specifically for soil conservation, the main measures for which were terracing and crop rotation.

It was not to combat soil erosion, but in an earlier tradition to concentrate population in areas free from sleeping sickness, that the largest settlement scheme in West Africa in the latter 1930s was designed. The Anchau Rural Development and Settlement Scheme in Nigeria[3] achieved the movement and housing of some 5,000 people among a total population of 50,000 affected by the measures. £95,000 was made available as a grant but, significantly for the future, on condition that an essential part of the scheme should be economic development. In the event, this economic development was limited to the provision of water supplies and various forms of propaganda. But elsewhere in Nigeria, at Daudawa, agricultural innovation in cash crops was the core of a settlement scheme which by 1943 supported 42 settlers through a cattle and cotton economy.[4]

[1] Quoted in Colin Maher, 'Soil Erosion and Land Utilization in the Embu Reserve' (mimeo), Vol. 1, p. 143.
[2] Ibid., Vol. 2, p. 67.
[3] T. A. M. Nash, *The Anchau Rural Development and Settlement Scheme*, H.M.S.O., London, 1948.
[4] C. B. Taylor, 'An Experiment in Land Settlement', *Tropical Agriculture*, Vol. 20, No. 4, April 1943, pp. 71–73.

In the early 1940s, in Nigeria at least, settlement was coming to be regarded as an opportunity for effecting change in economy and way of life.

It was not until the decade following the Second World War that settlement schemes in anglophone Africa enjoyed their first heyday. More capital was available. The Colonial Development and Welfare Act of 1940 had provided for the expenditure of £5 million a year in colonial territories,[1] but wartime pre-occupations and staff shortages had made it difficult or impossible to launch schemes. The 1945 Colonial Development and Welfare Act increased the sums to be contributed by the United Kingdom to £120 million in a ten-year period,[2] and this access of capital coincided with improvements in staffing positions and with a sense that new and more positive policies for underdeveloped colonies were required. In these circumstances, settlement schemes were more practicable than before. Those that were implemented varied in origins, purposes and organization, but can without excessive distortion be separated into three main streams, each associated with a different set of official ideas.

The first stream of settlement schemes had its source in attitudes of conservation. The depth of pessimism that could be felt about the prospects for African farmers is shown by a despatch from the Governor of Kenya in 1946:

Thirty years ago it seemed to us that the introduction of economic crops to be grown by African peasants afforded an opportunity of effective and permanent improvement in the standard of living and the production of wealth without a radical change in the methods of agriculture. Now I believe firmly that research will disclose that primary production by African peasants in the manner in which it has been hitherto developed is already on the decline, and that in fact, far from there being any substantial increase, populations working under this system are going to find increasing difficulty in supporting themselves even at the present level.[3]

Pessimism and fear were based upon ignorance not only of the potential human carrying capacity of tropical soils but also of the rates of growth of African populations about which in East Africa it was said that adequate statistics were not available.[4] Economically, socially and politically the position appeared unstable, and was

---

[1] Niculescu, op. cit., p. 62.

[2] Ibid., p. 62.

[3] Despatch No. 44 of 1946 from Sir Philip Mitchell, quoted in William Allan, *The African Husbandman*, Oliver and Boyd, Edinburgh, 1965, p. 390.

[4] Despatch No. 193 of 1951 from the Governor of Kenya, in *Land and Population in East Africa*, Colonial No. 290, H.M.S.O., London, 1952, p. 3.

taken particularly seriously in Kenya where African land hunger was regarded as a political threat to the areas settled by Europeans. The earlier concern with soils was now partly redirected towards people.

In most territories, settlement schemes were seen as one means of attack on the combined problems of erosion and overpopulation. Such schemes can be regarded as responses to ecological imbalances, as official attempts to control and stabilise. They were designed to attract or move people from areas of dense population out on to unoccupied land, thus reducing the pressures on the land from which they originated. Among the best documented are the Sukumaland Development Scheme in Tanganyika,[1] the Shendam Resettlement Scheme in Northern Nigeria,[2] the Kigezi Resettlements in Uganda,[3] the resettlements of Northern Rhodesia,[4] and a number of resettlement schemes in Kenya of which the Makueni Settlement[5] was the first and best known. In degree of planning and control they varied. The Kigezi Resettlements were little more than assisted migration with minimal services and resulted in little visible change when compared with areas of traditional farming; at Makueni holdings were laid out, agricultural practices controlled, and from 1957 a settlement fee charged. Although water supplies, roads and even bush clearance were sometimes provided for settlers, a high degree of self-help on their part was generally expected, and in time the area occupied became part of the normal administration of the country.

Approaches to overpopulation varied according to the demographic and political conditions of the different countries. In Nigeria, Uganda and Tanganyika, uncultivated but fertile land was relatively abundant and there were few political obstacles to new settlement. In these countries, therefore, overpopulation could be tackled largely by expanding the area under cultivation, whether by spontaneous or by controlled settlement. In Southern Rhodesia and Kenya, in contrast, European settlement had reduced the land available for African use, exacerbating overpopulation problems and at the same time creating a political situation in which African expan-

[1] Donald W. Malcolm, *Sukumaland, an African People and their Country, a Study of Land Use in Tanganyika*, Oxford University Press, 1953.
[2] E. O. W. Hunt, *An Experiment in Resettlement*, Government Printer, Kaduna, 1951; and William B. Schwab, *An Experiment in Resettlement in Northern Nigeria*, Haverford College, Haverford, 1954.
[3] Purseglove, op. cit.; E. S. Katarikawe, 'Agricultural Aspects of the Kiga Resettlement Programme in Western Uganda' (mimeo), Makerere University College, Faculty of Agriculture, R.D.R. Paper No. 11 (undated).
[4] George Kay, *Changing Patterns of Settlement and Land Use in the Eastern Province of Northern Rhodesia*, University of Hull Publications, 1965, and Allan, op. cit., pp. 446–452.
[5] *African Land Development in Kenya 1946–1962*, op. cit., pp. 40–46.

sion was regarded as a threat to security. The official responses in these two countries were remarkably similar in two respects. In the first place, extensive reorganization of African land tenure was undertaken. In Southern Rhodesia the operation known as land centralization, carried out under the Native Land Husbandry Act of 1951, while not strictly a settlement scheme, involved resiting homesteads in lines, standardization of land holdings, limitation of stock and a greater control of agricultural activities.[1] In the Kikuyu areas of Kenya the programme known as land consolidation, carried out during and after the Emergency of the 1950s, similarly involved a rationalization of land tenure, in this case by gathering together scattered fragments and re-allocating them as individual holdings; the programme also entailed a quasi-settlement operation as farmers moved out of the villages into which they had been collected for security reasons and established themselves on their new farms. Both land centralization in Southern Rhodesia and land consolidation in Kenya were justified in terms of conservation and agricultural development, but in both cases there were overtones of confining and controlling African farmers in their own areas.

The other similarity of official response to overpopulation in Southern Rhodesia and Kenya was irrigation settlement. Here the rationale of containing and controlling a political threat was more explicit. The Sabi Valley Irrigation Projects in Southern Rhodesia were greatly extended in the decade following the Second World War in order to resettle Africans displaced from the areas reserved for European farming.[2] Irrigation in Kenya was also designed to have a politically stabilizing effect. The Kikuyu detainee labour made available by the Emergency, especially from 1953 to 1956, was used to start three schemes—on the Perkerra river, on the Tana river, and at Mwea-Tebere—of which the latter was intended to provide settlement for displaced Kikuyu who might otherwise have been considered a security and political danger.

The second stream of settlement schemes in the post-war period was associated with the group or co-operative farming movement. These schemes were encouraged by supposed experience outside Africa, by ideology, and by what were believed to be technical considerations. The supposed experience was supplied by the West Indies and Palestine. Of the West Indies a Fabian Colonial Bureau Report stated that 'Land settlement facilitates the organization of individual farmers into co-operative units which can enjoy the

[1] C. Kingsley Garbett, 'The Land Husbandry Act of Southern Rhodesia', in Daniel Biebuyck, ed., *African Agrarian Systems*, Oxford University Press, 1963, pp. 185–199; and Yudelman, op. cit., pp. 115–129.
[2] Roder, op. cit., pp. 115–116.

advantages of large-scale agriculture'.[1] The co-operative settlements of Palestine were remarked upon by the Colonial Adviser on Agriculture, Sir Frank Stockdale, at the end of the 1938 Conference of Colonial Directors of Agriculture,[2] but relatively little appears to have been known about them except that they were considered successful.[3] After the Second World War this belief that co-operative forms of agriculture had been found to work elsewhere was reinforced by the ideological enthusiasm of the Labour Party in Britain for the co-operative movement. The Secretary of State for the Colonies in 1946, in asking Colonial Governments to extend the co-operative movement,[4] contributed to an atmosphere in which the idea of group farming became fashionable and was much discussed. Andrew Cohen reported of a summer conference on agricultural development in Africa, held in 1949, that:

> we talked a lot about 'group farming' and were, I think, generally agreed that group farming provided an organization that could help a very great deal in securing better farming practices and the necessary infusion of capital. We were reluctant to generalize about its practicability in different areas, but urged extensive experimentation in introducing systems of this kind.[5]

In the colonies themselves the movement was strengthened by technical arguments used by some officials. Pessimistic as they were about the prospects of developing modern agriculture on fragmented smallholdings, and aware perhaps rather imprecisely of traditional communal institutions, some agriculturalists argued that economies of scale and of level of technology could be introduced on land under inefficient small-scale African cultivation through communal farming sharing centralized facilities. The force of these beliefs is suggested by a striking and dogmatic resolution passed by the agricultural officers in Kenya at a conference in 1947:

> The policy of the Department for the Native Lands should in general be based on encouraging co-operative efforts and organiza-

---

[1] *Report of the Agricultural Policy Committee of Trinidad and Tobago*, Part 1, 1943, p. 17, quoted in *Co-operation in the Colonies: a Report from the Special Committee of the Fabian Colonial Bureau*, George Allen and Unwin, 1945, p. 158.

[2] *Report and Proceedings of the Conference of the Colonial Directors of Agriculture, July 1938*, op. cit., p. 84.

[3] *Co-operation in the Colonies*, op. cit., p. 160.

[4] *The Co-operative Movement in the Colonies, Despatches dated 20 March 1946 and 23 April 1946 from the Secretary of State for the Colonies to the Colonial Governments*, Colonial No. 199, H.M.S.O., London, 1946.

[5] *Agricultural Development in Africa*, Colonial Office Summer Conference on African Administration, African No. 117, H.M.S.O., London, 1949, p. 6.

tion rather than individual holdings. It is considered that only by co-operative action can the land be properly utilized, and the living standards of the people and the productivity of the land be raised, and preserved. While this involves a change from the modern trend towards individualism, it is in accord with the former indigenous methods of land usage and social custom.[1]

The term 'group farm' came to be applied to a variety of forms of organization, some of which involved a redistribution of population and some of which did not. As Lord Hailey pointed out, 'group farming' was a term of convenience rather than of precision.[2] However, neither the group farms which required little mechanization, such as those in Nyanza Province in Kenya,[3] nor those in which there was a high degree of mechanization, such as the Gonja Development Scheme in the Gold Coast, were successful. In Kenya it was observed that there was a tendency among African farmers towards 'rampant individualism',[4] and it came to be realized that the problems of organization and motivation in any form of co-operative production were far greater than had been thought. With hindsight agriculturalists in Kenya later considered group farming to have been a 'bogus problem' and a red-herring.[5]

The third stream of settlement schemes sprang from a complex mixture of vision, interests, faith, and capital, all of which were metropolitan in origin. The vision was expressed by Strachey, later to be the Minister of Food with oversight of the Groundnut Scheme in Tanganyika, who wrote in his war diaries: 'Thus far it has been possible to produce major collective efforts for the purposes of war alone. What could not be done if an expedition . . . could be fitted out not in order to decide who should . . . develop Africa, but in order actually to develop Africa.'[6]

The metropolitan interests were in increasing the production of primary goods within the sterling area both to overcome the post-war dollar crisis and to increase the supply of fats and other food-stuffs for Britain. This intention was summarized in the terms of reference of the Colonial Primary Products Committee called into being in May 1947, which were: 'to review, commodity by com-

---

[1] Quoted in 'Planned Group Farming in Nyanza Province, Kenya', *Tropical Agriculture*, Vol. 27, July–December 1950, p. 154.

[2] Hailey (1956), op. cit., p. 911.

[3] A. L. Jolly, 'Group Farming', *Tropical Agriculture*, Vol. 27, July–December 1950.

[4] 'Planned Group Farming in Nyanza Province, Kenya', *Tropical Agriculture*, Vol. 27, July–December 1950, p. 157.

[5] *African Land Development in Kenya 1946–62*, op. cit., p. 7.

[6] Quoted in a review of Alan Wood, 'The Groundnut Affair', in *Tropical Agriculture*, Vol. 27, July–December 1950, p. 160.

modity, the possibility of increasing colonial production, having regard on the one hand to the interests of the colonial empire and, on the other hand, the present and prospective world needs and the desirability of increasing foreign exchange resources.'[1]

Given what now appears an extraordinarily naïve faith that mechanical cultivation of unoccupied land in underdeveloped countries would lead to large increases in production and economically viable projects, it is not surprising that the Overseas Food Corporation and the Colonial Development Corporation, both set up in 1948, were used to channel large amounts of capital into major agricultural enterprises in some African countries.

Of the projects that resulted, the Groundnut Scheme in Tanganyika is of course the best known, but it was not intended to achieve settlement[2] and was in effect a number of state farms. In contrast, some other projects for large-scale mechanization did include settlement among their aims. A proposal for a 'pilot' scheme to settle 6,000 families in South Busoga in Uganda at an estimated cost of £850,000 was never implemented,[3] but the Gonja Development Company in the Gold Coast[4] and the Niger Agricultural Project in Nigeria[5] were similarly ambitious schemes which were undertaken. Both failed: the Gonja Scheme, started in 1950, was wound up in 1956; the Niger Agricultural Project, started in 1948, was closed down in 1954, having established only 163 settlers by 1953. Both schemes were intended to clear unoccupied savannah, to introduce mechanical cultivation, to increase foodstuffs for local consumption and for export, and to establish village settlements. They were abandoned for many reasons, not least agricultural and mechanical shortcomings, and in the case of the Niger Agricultural Project at least, social and economic factors affecting the settlers.

Another virtual failure, which stands outside the main streams of settlement policies, was the Zande Scheme in southern Sudan.[6] This evolved from proposals for regional and largely self-sufficient economic developments. Separate programmes for cotton growing and for resettlement were amalgamated. Between 1946 and 1950 almost 50,000 families, virtually the whole of the Azande population

---

[1] *Colonial Primary Products Committee Interim Report*, Colonial No. 217, H.M.S.O., London, 1948, p. 4.
[2] See Alan Wood, *The Groundnut Affair*, Bodley Head, 1950, pp. 80 and 175. The 'model villages' were for labourers who were to be employed on the project, not for settlers.
[3] Belshaw, op. cit., p. 4. A much smaller operation was undertaken, and failed.
[4] Anyane, op. cit., pp. 169–171.
[5] Baldwin, op. cit., *passim*.
[6] Reining, op. cit., *passim*.

in southern Sudan, were resettled on geometrically organized holdings, and cotton cultivation was made compulsory. The Scheme was very unpopular with the Azande whose financial returns were low, since it was necessary to subsidize a relatively small but uneconomic processing plant.

Some other schemes were the incidental result of other development projects. In Northern and Southern Rhodesia the Kariba dam on the Zambezi displaced a population of 57,000 who were moved to new areas in 1959.[1] Radical change in agricultural system was not part of the resettlement programmes. Elsewhere, settlement schemes were sometimes hangovers from earlier agricultural projects. In the three areas of the Groundnut Scheme small settlement schemes were maintained by the Tanganyika Agricultural Corporation after the main undertaking had been abandoned: at Nachingwea a mechanized farming scheme, which included African tenant farming, was started; but the number of settlers fluctuated, and dropped from 126 in 1955/56 to only 19 in 1960/61.[2] At Kongwa a similar mechanized tenant Farming Settlement Scheme avoided collapse by changing its economy from cultivation to ranching;[3] and at Urambo, the third groundnut area, tobacco proved successful, with central mechanized services provided for smallholder settlers.[4] In Ghana a small scheme for the settlement of Fra Fra survived after the demise of the Gonja Development Company.[5] In Uganda another settlement scheme which could be regarded as a hangover from an earlier project was started in South Busoga in 1956, after an earlier mechanized tenant farming project had been abandoned having achieved the clearing of only 750 acres for an expenditure of £100,000.[6] Also in Uganda the Kigumba Settlement Scheme was launched in 1957 on the land formerly used by the Bunyoro Agricultural Company which had folded up the year before.[7]

In the later years of colonial rule in most of the territories under consideration, settlement schemes were discredited. The dismal record of the more ambitious projects was discouraging in itself, but even more important was the growing realization that it was easier

---

[1] Colson, op. cit., p. 1.
[2] Hans Ruthenberg, *Agricultural Development in Tanganyika*, Springer-Verlag, Berlin, 1964, p. 81. See also R. F. Lord, *Economic Aspects of Mechanized Farming at Nachingwea in Tanganyika*, H.M.S.O., London, 1963.
[3] Peter Rigby, 'Settlement Schemes at Kongwa and their Significance for Socio-Economic Development in Ugogo' (typescript), September 1963. Also Ruthenberg, op. cit., pp. 86–87.
[4] Ruthenberg, op. cit., pp. 82–86.
[5] Anyane, op. cit., p. 169.
[6] Illingworth, op. cit., p. 12.
[7] Simon Charsley, 'Kigumba Settlement: the Establishment of Pluralists' (typescript), 1967.

to maintain tropical soil fertility than had been supposed, and that the human carrying capacity of land was far greater than had been believed. The East African Royal Commission, set up to consider problems among which over-population was prominent, duly suggested resettlement as one of the measures to deal with it,[1] but already in the early 1950s it was becoming clear that other methods might be more effective. The increasing cultivation of cash crops by African farmers, the high prices of the Korean war period, and the effectiveness of manuring and rotation on smallholdings all contributed to this new understanding.[2] Of one area in Kenya it was reported in 1955 that:

> A few years ago . . . the economic unit below which a man and his family could not earn a decent living, was placed as high as 100 acres. It has now come down to 25 acres and may go lower still.[3]

In these circumstances, many earlier settlement schemes were either handed over to local councils, maintained as subsidized experiments, or abandoned. A more strictly economic view was taken of those that survived as organizations. In Rhodesia after 1956 a realization that irrigation projects were subsidised resulted in a slowing of irrigation expansion.[4] In Uganda in 1962 the resettlement vote was cut from £50,000 to £10,000 a year,[5] and the Kigumba and South Busoga Schemes were more or less abandoned, the former having proved ineffective as a tsetse barrier, and the latter partly because of problems of land ownership. In Zambia in 1964 it was decided to bring to a close a Peasant Farming Scheme which had been introduced some sixteen years earlier.[6] In Kenya, the African Land Development Organization (ALDEV), which had initiated a number of settlement schemes, reflected changing official attitudes by shifting from grants to loans as its dominant form of financing.[7] Examples could be multiplied. Agricultural development programmes moved

---

[1] *Report of the East African Royal Commission 1953/55*, Cmd. 9475, H.M.S.O., London, 1955, p. 38.

[2] *African Land Development in Kenya 1946–1962*, op. cit., contains a discussion of the changing perceptions of the human carrying capacity of the land in Kenya at this time and (p. 7) 'the obsession with overpopulation and the need for more land' which was eventually dispelled.

[3] *Notes on Some Agricultural Development Schemes in British Colonial Territories*, op. cit., p. 28. The area was the Lambwe Valley, in Nyanza Province.

[4] Roder, op. cit., p. 198.

[5] Illingworth, op. cit., p. 159.

[6] *Annual Report of the Ministry of Agriculture, 1964*, Lusaka, Government Printer, 1965, p. 10.

[7] *African Land Development in Kenya 1946–62*, op. cit., p. 14.

away from settlement, and staff and finance were concentrated more on piecemeal improvement through agricultural extension work, especially the introduction of cash crops, credit schemes, and marketing co-operatives for smallholders. An observer might have supposed that the days of settlement schemes had passed.

Settlement schemes in independent[1] states

The decline in initiating settlement schemes which preceded independence proved only temporary. The problems, opportunities and aspirations which became evident in the years immediately before and after independence gave rise to a new generation of settlement projects. Some of these were forced on governments by populations displaced by political changes. The resettlement of southern Sudanese in Uganda, and of people from Rwanda in Uganda and Tanzania are examples, and similar operations were carried out elsewhere in Africa. Typically these refugee resettlements were carried out like the early sleeping sickness moves, with the intention of solving the human problem as quickly as possible and with a minimum of expenditure by enabling the displaced persons to become economically self-supporting. In these cases the completion of settlement was often what mattered most to officials, and there was little or nothing that could be described an on-going scheme. From 1966, however, largely as a result of initiatives by Oxfam a more positive view was being taken in Eastern Africa of the opportunity and need to follow through resettlement of refugees with programmes for agricultural development.[2] Other settlement schemes were more ambitious and diverse. In the colonial period a certain uniformity of origin and form derived from similar reactions of similar officials to similar ecological changes, and there was in addition a community of doctrine, information, and even staff between different colonies. In independence, in contrast, although there were continuities with previous practices, new conditions generated new and varied programmes for settlement. Political leaders were often unaware of the lessons of the failures of many colonial schemes, and were anxious to move away from the piecemeal approaches to agricultural development which had become

[1] The word 'independent' is adopted for convenience, since it applies legally to seven of the eight countries considered. The usage should not, however, be taken as implying any form of recognition of Rhodesia's Unilateral Declaration of Independence in 1965. The years in which independence was achieved were: Sudan—1956; Ghana—1957; Nigeria—1960; Tanganyika—1961; Uganda—1962; Kenya—1963; Zambia—1964.

[2] As outlined in T. F. Betts (Field Director, Oxfam), 'Refugees in Eastern Africa, a Comparative S tudy' (mimeo), Nairobi, 6 May 1966.

orthodox policy in the twilight of colonialism. In these circumstances, bold plans and fresh approaches were welcomed. The dissimilarity of the programmes that resulted was related to the variety of problems the schemes were designed to tackle, the heterogeneous foreign influences that conditioned them, and the very different national styles and ideologies that began to emerge.

The settlement programmes of independence arose out of different types of problems, and had different formal purposes. In Ghana, the Volta dam lake displaced a population which the Government evacuated and resettled. In both Western and Eastern Nigeria, agricultural settlement programmes were intended to counteract the drift of school-leavers to the towns by means of creating modern agricultural organizations and making rural life more attractive. In the Sudan the process of creating a nation out of desert was continued in the Gezira triangle, and the Halfawis made homeless by the rising waters of Lake Nasser were re-established at Khasm-el-Girba. Kenya, faced with widespread demand for the free distribution of European farms to landless Africans, responded with the Million-Acre Settlement Scheme which bought out Europeans and settled Africans in their place. Uganda, concerned to bring larger areas into cultivation and to achieve higher productivity, initiated a programme of mechanized group farming which sometimes took the form of settlement. Tanzania, in the interests of nation-building, sought to bring people together into villages where they could enjoy higher standards of services and the advantages of co-operation in agriculture. Rhodesia, in the Native Purchase Area Settlements, does not appear to have been doing more than provide a safety valve through which some of the more enterprising African farmers could obtain land. Zambia was concerned mainly with making the rural sector more attractive and profitable. In no two countries were the problems which gave rise to organized settlement programmes quite the same. Nor, apart from the universal intention to increase production, was there uniformity in the purposes they were intended to fulfil.

A further source of diversity was foreign aid and advice. There was a conjunction of opportunities and interests between donor and adviser organizations on the one hand and recipient countries on the other. All these wanted projects which would be visible and identifiable: donors and advisers so that they could show results for their aid or advice; and recipient governments so that they could show their people that national problems were being tackled vigorously. It is not surprising that high capital projects involving high imports from donor countries tended to be preferred, leading to such absurdities as the import into Tanzania from the U.S.A. of metal frames for settlers' houses, despite an abundance of local materials. At the same time, development in the immediate post-independence period was strongly influenced by the universal move towards national planning. Since indigenous economists were one of the scarcest forms of manpower, most national plans were largely drawn up by foreign advisers, and these came from many different countries and brought with them many different ideas. Since development planning placed a premium on projects and programmes, it was almost inevitable that these different ideas should influence the forms of organization proposed for government-sponsored development. Settlement schemes were no exception. Except for the Sudan, where it was known to work well, the Gezira model faded into the background. Its place was partly taken by a number of variations on the theme of the Israeli *moshav*, involving small individual farms with centralized services for cultivation, marketing and social welfare. The farm settlement programmes of Western and Eastern Nigeria were explicitly based upon the *moshav*[1] which also, through the recommendations of Israeli advisers,[2] influenced the form of organization of the Tanzania Pilot Village Settlements.[3] It is intriguing to speculate what other forms of settlement organization might have been adopted had it not been in the interests of Israel at the time to use technical assistance as a means of establishing good relations with black African countries as a counterbalance to the Arab League.

The nature of settlement was also influenced by individual personalities, both those who organized and managed government schemes and those who started schemes of a semi-private nature. Among the latter, the Nyakashaka Resettlement Scheme in Uganda and the Ruvuma Development Association (R.D.A.) in Tanzania had in common that they were led by outstanding persons. Nyakashaka, started in 1963 by Stephen Carr, formerly a missionary in southern Sudan, had by 1966 settled 87 school-leavers on six-acre plots. Tea had been planted and the scheme was steadily expanding.[4] The R.D.A., which had its beginnings in 1960, was inspired by an Englishman, Ralph Ibbott, and a Tanganyikan, Ntimbanjayo Millinga. By 1966 it had attracted 250 families into 12 villages where

[1] Mordechai E. Kreinin, *Israel and Africa: a Study in Technical Co-operation*, Praeger, New York, 1964, pp. 45 and 48–55.

[2] For example, Benjamin Kaplan, *New Settlements and Agricultural Development in Tanganyika: Report and Recommendations to the Government of Tanganyika Resulting from a Study Mission Sponsored by the Government of Israel*, State of Israel, Ministry of Foreign Affairs, Department of International Co-operation, Jerusalem, 1961.

[3] Apthorpe, op. cit., p. 2.

[4] 'Caroline Hutton', Case Study of Nyakashaka Resettlement Scheme, a Preliminary Account (mimes) East African Institute of Social Research Sociology-Anthropology Workshop, June 1966.

they were practising a communal form of agriculture.[1] Although very different in organization, these two enterprises were similar in having strong ideological content, Christian and socialist respectively, and in finding a favourable political reception in the environment of independence.

The most pervasive influence on the form and purpose of settlement schemes was, however, emergent national style, reinforced in some cases by ideology. These different styles were themselves related to different physical, economic, social, and political conditions.[2] In this connection it is important not to be misled by the apparent similarities between countries in the stress on socialist values and co-operatives, for the meanings of socialism were many, and the scope of co-operatives ranged from limited *ad hoc* marketing to comprehensive collective production. Settlement schemes, as positive acts which brought land and people together in new combinations, offered opportunities for both planners and politicians to express their ideas about society and about the nation they sought to create. Indeed, settlement schemes could be all things to all men, embodiments of diverse Utopian aspirations. As Apthorpe has put it, they were 'experiments in nation-building in miniature'.[3] The sort of nation that was being built and the sort of people it was to benefit were different from country to country.

Continuity with past policies was most marked, though with very different results, in Rhodesia and the Sudan. In Rhodesia, settlement in the 1960s followed the existing pattern. Under the Land Apportionment Act of 1930 the Government was to provide for the purchase of land by Africans in the Native Purchase Areas. Settlement schemes, of which that at Chesa[4] is an example, provided for survey, land purchase, and agricultural extension services, but little else, in a style similar to the controlled migrations of the colonial period elsewhere. The result was a series of smallholdings with minimal government controls. In the Sudan, in contrast, post-independence settlement schemes followed the Gezira pattern in effecting a transformation of agricultural practices. The Managil

---

[1] Interviews (1966) with Ralph Ibbott and Ntimbanjayo Millinga. See also Ralph Ibbott, 'The Ruvuma Development Association', *Mbioni*, Vol. 3, No. 2, July 1966, pp. 3–43, and G. L. Cunningham, 'The Ruvuma Development Association, an Independent Critique', ibid., pp. 44–55.

[2] The emergent national cultures of Kenya, Uganda and Tanzania and their relationships with different land tenure systems and types of settlement scheme are discussed in Aaron Segal, 'The Politics of Land in East Africa' (mimeo), paper to the conference of the American African Studies Association, 1966.

[3] Apthorpe, op. cit., p. 22.

[4] Roger Woods, 'The Dynamics of Land Settlement: Pointers from a Rhodesian Land Settlement Scheme' (mimeo), Dar es Salaam, December 1966, describes the Chesa Native Purchase Area Settlements.

Extension, which was progressively brought into production from 1958 to 1962, adopted many of the methods that had been found to work on the Gezira main area, the principal alterations being a smaller tenancy (15 feddans[1] instead of 40), more intensive cultivation, and a closer planning of villages.[2] The resettlement at Khasm-el-Girba of the Halfawis displaced by Lake Nasser also involved a radical economic change, as the evacuees gave up the town life of Wadi Halfa for farming by controlled irrigation.[3] In keeping with the previous standard of living of the Halfawis, and the welfare philosophies that were common among independent governments, high standard housing was provided. While this housing was an innovation, the adoption of large-scale irrigation for settlement was in the long accepted tradition of Sudanese development.

In Ghana the evacuation of the population of some 80,000 from the basin of the Volta dam, which took place mainly in 1964, was also affected by national experience and ideology.[4] The difficulties earlier experienced in resettling the fishing village of Tema[5] to make way for the new port encouraged a benevolent and permissive approach. The quasi-urban values of West Africa were combined with socialist forms, if not realities, of organization in collecting the displaced population, originally in over 700 villages and hamlets, into 52 new towns in which relatively high standard housing was provided. Agriculturalists seized upon the opportunity to propose mechanized agriculture for the new towns, but the agricultural programme fell far behind schedule so that there was a danger that some of the resettlements would become ghost towns through absenteeism as evacuees migrated in search of a better livelihood. The shortcomings of the agricultural programme were but one example of the failure of ambitious agricultural projects with overtones of socialism initiated in Nkrumah's Ghana.

The principal settlement programmes launched in Southern Nigeria and in Uganda show some intriguing differences and similarities. They differed in formal purpose and in methods of implementation. In Southern Nigeria the official justification was that capital-intensive co-operative farm settlements could be used to settle school-leavers on the land and to demonstrate that rural life

---

[1] A 'feddan' is 1·038 acres.
[2] Shaw, op. cit., especially pp. 11–20.
[3] Thornton and Wynn, op. cit.
[4] A major source is *Volta Resettlement Symposium Papers* (papers read at the Volta Resettlement Symposium held in Kumasi 23–27 March 1965) (mimeo). Volta River Authority, Accra, and Faculty of Architecture, University of Science and Technology, Kumasi, 1965.
[5] Described in G. W. Amarteifio, D. A. P. Butcher and David Whitham, *Tema Manhean, a Study of Resettlement*, Ghana Universities Press, Accra, 1966.

could be at least as profitable and attractive as urban. In the Western Region of Nigeria, thirteen such settlements, of which the Ilora Farm Settlement was one,[1] were started in 1960. In the Eastern Region, six farm settlements were established in a programme launched in 1962. These later Eastern Nigerian settlements differed from the earlier Western ones in the selection of somewhat older settlers, training them more before settlement, and allocating smaller plots, but both programmes relied on similar agricultural organization. In Uganda, however, the group farms[2] presented a wider range of organization and were meant to be based upon existing strong co-operative societies. In practice they varied from non-settlement tractor hire services to full settlement schemes and in the official rationale emphasis was placed on increasing agricultural production.

Notwithstanding these contrasts, the Nigerian and Ugandan programmes had in common two significant features: both involved the introduction of mechanization with a high degree of subsidy; and both established schemes with a wide geographical scatter. It can scarcely be considered a coincidence that at the time both Nigeria and Uganda had competitive political systems in which the governing party had to maintain a majority in parliament and might expect to have to retain its position through popular elections. Indeed, the reasons for mechanization and the physical dispersal of scheme, so frequently opposed by advisers, appear neither agricultural nor economic, but political. In Western Nigeria, for instance, the recommendation of Israeli advisers that *moshavim* should be established in three compact complexes so that they could share central services was not accepted, and instead the individual settlements were widely separated as independent units. In Uganda, group farms were similarly dispersed, and distributed as forms of patronage and inducement in politically strategic areas, especially in the north of the country.[3] In both Western Nigeria and Uganda the scattering and subsidising of agricultural projects appeared to result largely from the need of the ruling parties to secure and maintain support by dispensing visible and spectacular rewards in rural areas.

In their settlement programmes Tanzania and Kenya present a marked contrast to other countries and to each other. The pro-

---

[1] The Ilora Farm Settlement is described in Oladejo O. Okediji, 'Some Socio-Cultural Problems in the Western Nigeria Land Settlement Scheme: a Case Study', *Nigerian Journal of Economic and Social Studies*, Vol. 7, No. 3, November 1965, pp. 301–310.

[2] A Uganda group farm is described in Simon Charsley, 'Group Farming in Bunyoro', *East African Institute of Social Research Conference Papers*, January 1966, and 'The Profitability of a Group Farm', *Makerere Institute of Social Research Conference Papers, January 1967*, No. 405.

[3] This point is elaborated in Segal, op. cit., pp. 6–8.

grammes and the philosophies which lay behind them can be understood as arising from different demographic and economic situations. In Tanzania, with a scattered population and relatively abundant land, nation-building and development were sought through the collection of people into villages. In 1962, President Nyerere put it that: 'Before we can bring any of the benefits of modern development to the farmers of Tanganyika, the very first step is to make it possible to start living in village communities.'[1]

In harmony with this policy, the Tanzanian idea of African socialism and *ujamaa* valued communal activities and co-operation.[2] The national campaign for village settlement that followed the President's announcement took many forms: TANU Youth League settlements, almost all of which were short-lived; clusterings of homesteads into larger residential units in some parts of the country; and eight high capital, closely managed, Pilot Village Settlements, which were controlled by a Rural Settlement Commission.[3] Of these pilot schemes, that at Upper Kitete may be typical in size, having by mid-1965 settled 100 farmers in three villages, with an economy based on communal mechanized cultivation of wheat and livestock herding. The Rural Settlement Commission also took over nine schemes formerly managed by the Tanganyika Agricultural Corporation, including among others three—at Kongwa, Nachingwea, and Urambo—which were hangovers from the Groundnut Scheme, and Kiwere, a settlement based upon mechanized cultivation of tobacco. The pilot programme was suspended in 1966, however, when it had become clear that the village settlements were uneconomic compared with other investments and were turning the settlers into a specially privileged group with attitudes of dependence on government, in conflict with the emergent national ethos of equality and self-reliance.[4]

In Kenya the acute perception of land shortage, especially among Kikuyu and Luo, and the widespread sensitivity about land rights largely resulting from European occupation of the so-called White Highlands, led to a far more individualist approach to settlement

---

[1] *Tanganyika Parliamentary Debates, National Assembly Official Report*, 1st Session, 10 December 1962 to 16 February 1963, p. 7.

[2] Julius K. Nyerere, *Freedom and Unity: Uhuru na Umoja, a Selection from the Writings and Speeches 1952–65*, Oxford University Press, 1967, especially pp. 162–171, '*Ujamaa*—the Basis of African Socialism'.

[3] The eight Pilot Village Settlements were Amani, Bwakira Chini, Kabuku, Kerege, Kingorogundwa, Mlale, Rwamkoma, and Upper Kitete.

[4] The development of village settlement policies in Tanzania up to December 1966 is summarized in Garry Thomas, 'The Transformation Approach at a Tanzania Village Settlement', *Makerere Institute of Social Research Conference Papers, January 1967*, No. 427, pp. 1–5. See also the bibliography for other references.

with family autonomy rather than communal co-operation as the ideal condition that was sought. The understanding of African socialism put forward in Kenya[1] can be interpreted in this light as a rationalization of the interests of a rural bourgeoisie, of a nation of smallholders or would-be smallholders. In this spirit the (African) Director of Settlement in Kenya in his report for 1965-1966 stressed the need for his organization to commit itself to success based upon 'the conviction that smallholders, whether in Settlement or not, will be the mainstay of this agricultural country'.[2] Most of the 135[3] Settlements of the Million-Acre Scheme[4] were designed to provide individual plots for settlers, thus removing political discontent by satisfying the desire of landless men for their own pieces of land. Whereas Tanzania sought to aggregate population for nation-building, Kenya sought to disperse it for security. Whereas Tanzania sought in the Pilot Village Settlements to transform the agricultural economy, Kenya in the Million-Acre Scheme adopted a more gradualist approach, with greater devolution to the individual farmer and a concentration of effort on extension services and creating co-operatives for marketing. There were, however, two notable exceptions. First, on the 130,000 acres of the O1 Kalou Salient, which was suitable only for large-scale wheat and livestock production, co-operative farming was undertaken with high capitalization and central mechanical services. Second, at Muhuroni, mechanized sugar production on a quasi-estate system was started with settlers responsible for contiguous plots of sugar cane. In both these cases concessions were made to the settlers in allowing them farming areas on which they could grow their own foodstuffs, and mechanization was adopted only because it was the most economical form of organization for the land and crops in question.

In Zambia the post-independence approach to settlement schemes was varied and cautious. As elsewhere the purposes of schemes were related to national problems. The relief of rural overcrowding and the protection of soil were in a long-established tradition, but the aggregation of scattered population and the provision of work for the unemployed were part of a fresh attempt (characterized as 'back to the land') to correct the imbalance between the relative attractiveness to workers of urban and rural life. Moreover, a further aim was to diversify the national economy, vulnerable through its heavy reliance on copper. In early 1967 Zambia was embarking upon two

---

[1] *African Socialism and its Application to Planning in Kenya*, Government Printer, Nairobi, 1965.
[2] *Department of Settlement, Kenya, Annual Report 1965–1966*, Nairobi, 1966, p. 1.
[3] Ibid., p. 53.
[4] See the bibliography for references to this Scheme.

pilot settlement projects in each province, one for village re-grouping and one for the establishment of a rural service centre, besides undertaking small-scale schemes for the settlement of youths from the Zambia Youth Service. Although over £2 million had been allocated for settlement in the 1966-1967 financial year, the Zambian approach appeared more prudent and circumspect than in most other countries.

Although in independence each country developed its own form of settlement, each also inherited a legacy of previous settlement schemes. Many of these, especially those which had brought about major changes in land use or agricultural system, continued as visible land and organizational units. These could be seen to have an existence of their own. Among those that could be identified in Kenya were some survivors of the ALDEV settlement schemes with relatively simple agricultural organizations, and the three more complex irrigation schemes of Perkerra, Tana, and Mwea.[1]

In any study of settlement schemes it may be supposed, *a priori*, that the examination of a scheme will be more rewarding the more clearly identifiable it is, the longer it has been in existence, and the more specialized its organization. The Mwea Irrigation Settlement goes far to meet these conditions. It had, by 1967, clear physical, organizational, social and economic boundaries. It had been started in 1954 so that its history was long enough to display the process of formation and growth of a scheme. It had continued to exist through major political changes so that its record might reveal factors affecting the survival and prosperity of a scheme. Its bureaucracy of staff and its community of tenants were differentiated with specialized roles for different groups. Being relatively large and complex it could be expected to contain a full inventory of the components and relationships to be found in a scheme, which might be useful in working towards a typology of schemes. Furthermore, being regarded as successful it might reveal clues for the successful planning and implementation of other projects. The study of its history and organization which follows in Part II should, therefore, assist an understanding of settlement schemes in general and also lead to conclusions with practical implications.

[1] For a description of the Perkerra and Mwea Schemes see de Wilde *et al.*, op. cit., Vol. II, pp. 221–241. See also the bibliography.

Part II

The Mwea Scheme:
An Administrative Case History

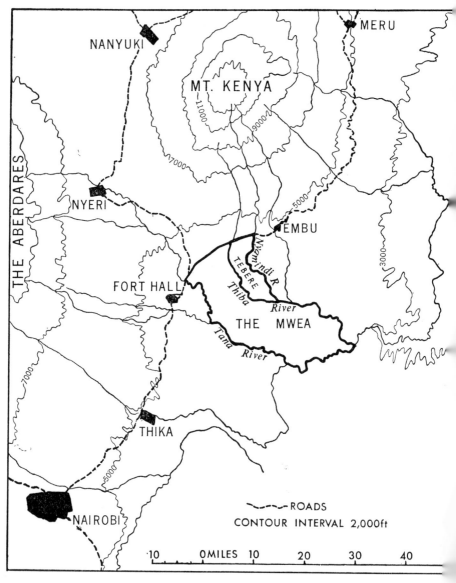

THE POSITION OF THE MWEA IN KENYA

CHAPTER 3

Environment and Origins

At the turn of the century the part of Kenya known as the Mwea[1] was largely unoccupied territory. Human settlement was discouraged by tsetse fly, low rainfall, dangerous game animals, and intermittent raiding by Masai. It was an inhospitable area of bush and high grass with permanent water only in a few rivers. Indeed, the word 'mwea' means an extensive, desolate region.

By the end of 1967 much of the Mwea had been settled. The more fertile areas were being cultivated and the drier areas grazed. The bush, high grass and tsetse fly had been reduced and almost all the game driven out. But by far the most conspicuous change was the presence of a system of irrigation which distributed water from two rivers to nearly 7,000 level one-acre fields. This system was operated by a staff of some 150 who controlled water distribution, supplied mechanical services, and supervised agricultural operations. It supported over 1,700 tenants and their families and produced about 14,000 tons of rice paddy a year. The Mwea Irrigation Settlement, as it was known, was regarded as a successful development scheme and was steadily expanding in area.

How and why did these changes come about? Who were the people involved? What organizations took part? And how were these people and organizations related in the processes of change which created the Scheme?

The land and the people

The story of the origins and development of the Mwea Irrigation Settlement is a drama of competitive colonization, using that word mainly in its biological and geographical sense. The stage for this drama, the Mwea,[2] is an area of some 215 square miles with its centre 60 miles north-east of Nairobi, in shape like an enormous flat

[1] For the position of the Mwea in Kenya see the map on p. 42.
[2] Conventions have varied about the exact boundaries of the Mwea. The most commonly accepted official view, at least until the early 1950s, was that it was the area bounded by the Embu-Sagana road and the Tana and Thiba rivers. In common usage, as more and more of the upper Mwea has been settled, it has tended to be excluded from what is meant by 'the Mwea'.

right footprint, with the heel at about 4,000 feet above sea level on the slopes of the Mount Kenya massif and the big toe some 700 feet lower where the Tana and Thiba rivers flow together. Very crudely, the Mwea can be described in terms of two ecological zones,[1] the upper Mwea, two-fifths of the area, corresponding roughly with the heel of the foot; and the lower Mwea, the remaining three-fifths. Although the irrigation scheme is in the upper Mwea its origins are bound up with the lower Mwea as well.

The upper Mwea has a gently undulating landscape based upon volcanic lava flows which spread out from Mount Kenya and which have been dissected by permanent rivers into broad ridges and shallow valleys. The soils of the ridges are mainly reddish volcanic loams, the 'red soils', while the flatter areas usually consist of heavy impervious clays, the 'black soils'. The mean annual rainfall, between 40 and 30 inches a year, permits some cultivation. With the notable exception of the 2,000 acre Nguka swamp, most of the original vegetation was grass woodland. To the east the contiguous area of Tebere, bounded by the Thiba and Nyamindi rivers and the Embu-Sagana road, is ecologically similar.

In contrast the lower Mwea is based upon older metamorphic rock with a more mature drainage system and steeper slopes. The main feature is a hilly ridge running between the Tana and Thiba rivers. The soils are lighter than in the upper Mwea and more easily erodable. A lower rainfall allows only speculative cultivation, and ranching is considered the correct land use. The vegetation is thornbush and grassland with some wild sisal in the drier areas.

Early in the century the Mwea was bounded by land occupied by five groups of people: the Kikuyu, Ndia, Gichugu, Mbere, and Kamba.[2] All five groups were Bantu, speaking languages that were

[1] Geological, climatic, soil and other ecological information is given in W. A. Fairburn, 'Geology of the Fort Hall Area', Geological Survey of Kenya Report No. 73, Kenya, Government Printer, 1965; Embu District Agricultural Gazeteer 1955/56 (mimeo); Upper Tana Catchment Water Resources Survey 1958-9, Report by Sir Alexander Gibb and Partners (Africa), Kenya Government, April 1959, especially pp. 40–41 for soils; and 'The Mwea', a report dated 8th January 1953, by a former District Agricultural Officer of Embu District, in Ministry of Agriculture file IRRIG/MWEA at folio 15A.

[2] See map on p. 45 for their relative positions at that time. The Ndia and Gichugu are usually considered to be subtribes of the Kikuyu, though more distinct than other subtribes. They will be referred to separately throughout, while other subtribes will be described collectively as Kikuyu, although that most concerned with the history of the Mwea was the Metume subtribe in Fort Hall District. Ethnographic descriptions may be found in John Middleton and Greet Kershaw, The Kikuyu and Kamba of Kenya, Ethnographic Survey of Africa edited by Daryll Forde, East Central Africa Part V, International African Institute, London, Revised Edition, 1965; and H. E. Lambert, Kikuyu Social and Political Institutions, Oxford University Press, 1956.

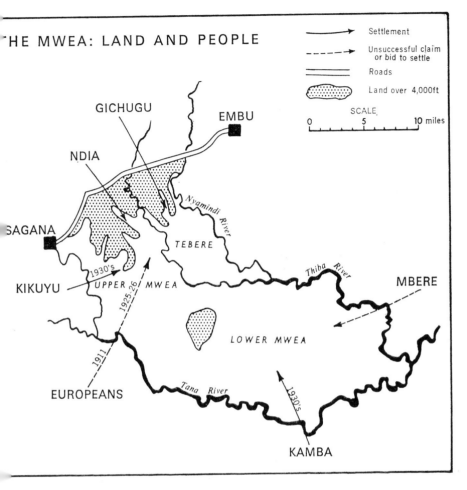

THE MWEA: LAND AND PEOPLE

to varying degrees mutually intelligible. They could inter-marry, had generally similar social institutions, lived typically on separate homesteads, and had similar economies depending upon shifting cultivation of food crops and the herding of cattle, sheep and goats. They were, however, also similar in their political autonomy so that they were potential rivals for land, whether for grazing or cultivation. Following the assertion of the *pax Britannica* throughout the area in the first decade of the century human and animal populations began to follow the classical pattern of expansion found in so many parts of Africa, and grazing and settlement began to spread outwards from the more fertile areas that were already occupied on to more marginal lands. Each tribal group sought an area into which to expand. The Gichugu looked mainly towards Tebere. The Mbere, the Kamba, the Kikuyu, and especially the Ndia looked towards the empty lands of the Mwea.

In the event, territorial claims and disputes were precipitated by the arrival of a sixth group, the European settlers, who in 1911 and again in the latter 1920s sought approval for farming enterprises in the upper Mwea near the Nguka swamp. The Ndia, under the leadership of Chief Njega, were able to reject these proposals, but the presence of European settlement on the south bank of the Tana and the loss of land by Kikuyu elsewhere made the threat of European acquisition of the Mwea credible, and strongly conditioned subsequent Ndia attitudes. Chief Njega was, and remained, deeply distrustful of real or imagined European designs. In 1936 when 700 acres of the Mwea were to be flooded by the dam of the Tana-Maragua Electric Power Station, and cash compensation was offered to the Embu Local Native Council he is reported to have said: 'I do not want to see the Council receive money instead of that land because people will then say that the Council have sold Mwea. . . . I would rather grant the land to the Company and take nothing in exchange, as we would grant land to any stranger who begged it of us, than lay ourselves open to the charge of selling Mwea.'[1]

Chief Njega died in 1948 without retracting an oath he had sworn never to be a party to any sale of the Mwea. The leaders of the Ndia remembered this and their suspicion of European and Government motives for interest in the land they regarded as theirs profoundly affected the course of development in the area.

At first Government took little interest in the Mwea. Being largely unoccupied it could be transferred easily from district to district, and before 1933 there were no less than four successive arrangements for its administration.[2] But government interest followed in the wake of tribal disputes. In 1933 Chief Kombo of the Mbere and Chief Njega of the Ndia submitted to the Kenya Land Commission[3] their rival claims to part of the Mwea. Chief Njega claimed that the Ndia had asserted their rights by knocking Mbere honey-barrels out

[1] Minutes of Embu L.N.C. meeting of 6 January 1936.
[2] *Kenya Land Commission: Evidence and Memoranda*, Vol. 1, Colonial No. 91, H.M.S.O., London, 1934, p. 530 (Memorandum from Mr J. W. Pease, District Commissioner, South Nyeri).
[3] Commonly known as the Carter Commission after its Chairman, Sir Morris Carter.

of trees with stones,[1] and Chief Njega was generally considered to have won his case. But at this time, as so often in the history of rural development, the emptiness of land attracted official attention. The Commission was looking for land that could be given to the Kikuyu in compensation for that wrongly settled earlier by Europeans, and came out with the recommendation that much of the Mwea should be used for that purpose.[2] While this reduced the danger that the Mwea would be settled by Europeans it was bitterly resented not only by the Kikuyu, whom it was meant to placate, but also by the Ndia who wished the area to be theirs and not an expansion area for others.

The Ndia, however, adapted to the immigration of other groups in ways which implied and even reinforced their claims. During the 1930s Kamba crossed the Tana and settled in the lower Mwea, and it was probably no coincidence that Chief Njega came to have a number of Kamba wives. More extensive and significant was the entry of Kikuyu from Fort Hall District into the upper Mwea. Although at this time there was little Ndia cultivation in the Mwea, these Kikuyu immigrants were made to recognize Ndia claims either by becoming tenants-at-will (*ahoi*) or by being born again into Ndia clans and becoming, effectively, Ndia themselves. In either case, by recognizing Ndia claims, they were making possible for the Ndia a sort of colonization by proxy.

Official responses to this Kikuyu immigration were mixed. The settlers introduced extensive farming using ox ploughs instead of hoes, and farmed for profit. To the Administration they were welcome as 'providing a leaven of progressive ideas to the less advanced natives of this district'[3] and by the 1940s their techniques were being imitated by the local people. But Agricolas[4] were alarmed at the effects of what they regarded as a destructive form of agriculture. As one of them wrote in 1938: '. . . these Kikuyu migrants are displaying the usual energy and enterprise together with the destructiveness and lack of foresight for which Kikuyu agriculture is both famed and notorious'.[5]

[1] *Kenya Land Commission: Evidence and Memoranda*, op. cit., Vol. 1. For Chief Kombo wa Munyiri's evidence see p. 251, and for Chief Njega wa Gioko's see p. 252.

[2] *Report of the Kenya Land Commission*, Colonial No. 4556, H.M.S.O., London, 1934, pp. 65–67 and 76–77.

[3] Embu District Annual Report for 1934, quoted in Maher, op. cit., p. 137.

[4] The term 'Agricola' is borrowed from Elspeth Huxley, *A New Earth: an Experiment in Colonialism*, Chatto and Windus, 1960, p. 24 and *passim*, and is used here to refer to officers of the Department of Agriculture, and later to include agricultural managers of ALDEV.

[5] Maher, op. cit., p. 136.

Both the Administration and the Agricolas agreed, however, that the position would be improved if the immigrants could be given greater security of tenure. To the Administration this would have the beneficial effect of settling the population more permanently, and giving some plausibility to the effectiveness of the Land Commission's recommendation that the area be used to compensate the Kikuyu. To the Agricolas it would have improved the chances of a farmer taking a long term view of the need to conserve soil and fertility. But official attempts to increase security of tenure for the immigrants were only moderately successful, and in the 1940s there was increasing hostility on the part of the Ndia to the Kikuyu *ahoi*, and discrimination against them. It appeared that some of the suspicion and hostility originally aroused by European attempts to settle in the Mwea had been partly redirected against the Kikuyu.

The change of attitude, from welcoming to resenting Kikuyu immigrants, can be understood in terms of ecological changes. As the human and animal population of the Ndia and Gichugu increased, and as the pressure on the land reduced the fallows of shifting cultivation, settlement and grazing were forced out of the higher rainfall areas above the Embu-Sagana road and down on to the Mwea plains where the Kikuyu *ahoi* had settled. Competition was aggravated by the pattern of land use. Since the black soils were too heavy to cultivate, and during the rains when the seasonal grazing mainly took place were boggy and gave cattle sore hooves, both cultivation and grazing were concentrated on lighter and more easily erodable red soils of the ridges. Kikuyu settlement on the red soils contributed to forcing Ndia and Gichugu stock further away from the parent homesteads above the road, thus denying milk to the people and manure to their farms. In these circumstances there was mounting resentment of the Kikuyu on the part of the Ndia and Gichugu, and a reluctance to grant them security of tenure.

As for the officials, the damage to land from overgrazing and what was called destructive cultivation evoked the classical responses of the soil conservation era. In 1937 Soil Erosion Rules[1] were piloted through the Local Native Council. Government officers followed behind the frontiers of grazing and settlement and increasingly penetrated the Mwea. In the lower Mwea large areas of land were closed to grazing in 1947 and successfully reconditioned by scratch ploughing and reseeding.[2] In the upper Mwea, however, the problems

---

[1] Embu L.N.C. Regulation No. 7 of 1937.
[2] The grazing history of this area is summarized in R. H. Brown, 'A Survey of the Grazing Schemes operating in Kenya', Department of Veterinary Services, 1959 (mimeo), p. 30.

were more intractable. During the 1940s the Ndia and Gichugu, remembering European bids to settle and responsive to the rising political temperature in the country, were increasingly suspicious of the European Government and hostile to the controls it sought to impose. As elsewhere in central Kenya, soil conservation became one of the foci of opposition.[1] The earlier struggles between the Ndia, Mbere and others for rights of land use were now overlaid by struggles between government and people over control of land use. These struggles can be seen as a continuation of ecological change. Government had in the first place established the conditions for increases in population which pushed grazing and cultivation out on to the marginal lands of the Mwea. As the previous balances in those areas were upset, government attention and controls followed as the next invasion of the environment, the next level of the succession, in attempts to stabilize what had been disturbed. 'Government', however, was not monolithic: to use it as a blanket term would be as misleading as to talk only of 'local people'. In much the same way as the local people were divided into groups, so government consisted of departments; but these differed from tribal groups in being organizations rather than societies, and in having specialized functions. Like the tribal groups these departments were also potential collaborators and competitors, and must be described if events are adequately to be understood. Only then will it be possible to appreciate how, out of this background, an irrigation project came to be born and why it evolved as it did.

Departments of government

The four departments most closely involved in the origins and development of the irrigation scheme were the Administration, the Department of Agriculture, the African Land Development Organization (ALDEV) and the Ministry of Works.

'The Administration', as a colloquial phrase, was used in Kenya as in other British Colonies in Africa to describe Provincial Administration in the field. For the latter 1940s and the 1950s which are considered here, however, it is important to recognize that administrative officers occupied positions of great influence not only in the field Provincial Administration but also in central government. At the centre the Chief Native Commissioner was senior to other civil servants, and as head of the Provincial Administration had special access to the Governor and was his principal adviser on

---

[1] Opposition to soil conservation in Kenya at this time is discussed more fully and generally in Carl G. Rosberg, Jr., and John Nottingham, *The Myth of 'Mau Mau': Nationalism in Kenya*, Frederick A. Praeger, New York, 1966, pp. 237–238.

African affairs. Technical departments, such as the Department of Agriculture and the Medical Department, though headed by senior technical officers (the Director of Agriculture, the Director of Medical Services, etc.) were administered and overseen by administrative officers at first as Secretaries, and later as Permanent Secretaries. Again, some of the Deputy Secretaries, Under Secretaries and Assistant Secretaries who worked to the Permanent Secretaries were also administrative officers. Perhaps most significantly the Treasury, which exercised controls over departmental financial allocations, was primarily staffed with officers from the Administration who did not hesitate to criticize and amend plans put up to them by technical departments. Further, since it was official policy that administrative officers should serve at different times both in the field and in central government, a certain coherence of view and sympathy of interests between field administrators and senior administrative officers in the centre was encouraged, and much communication took place informally on an 'old boy net'. Rapport and familiarity were reinforced by departmental and racial loyalties and also by the similar social and educational backgrounds of administrative officers.

In the field the primacy of the Administration can be understood historically. As the first government department to appear in an area, it assumed all the responsibilities of government: indeed it was the Government. Within each district an organization (typically of Chiefs in charge of locations, Headmen in charge of sublocations, and Tribal Police responsible for law and order) was set up as part of the Administration, providing a means of controlling local affairs and an effective basis for authority. The Administration thus had not only what may be described metaphorically as rights of first occupation, but was also the first government organization that extended to the grass roots. Associated with this dominant position which amounted to a sort of sovereignty were a number of proprietary attitudes common among field administrative officers. The three most important—towards land, people and roles respectively—will be considered briefly.

In the first place field administrative officers were closely identified with land units. As Lucy Mair has put it, the D.C. in Africa was 'a new kind of territorial chief'.[1] The P.C. in his province, the D.C. in his district, and the D.O. in his division were often peculiarly territorial in their attitudes and behaviour. They sometimes took a greater interest in the definition of 'their' boundaries than did the people whose interests the definition was supposed to serve. When they met on boundaries, their behaviour towards one another could

[1] Lucy Mair, *Primitive Government*, Penguin Books, 1962, p. 253.

be territorial in the ethological sense in the way they sited their camps, in display, and even in aggressive behaviour towards one another.

The second set of proprietary attitudes, towards the local people, was associated with what were regarded as the Administration's political functions. To the Administration the local people were a constituency,[1] a source of support that could be used against other organizations, and also a source of obligations. The Administration assumed a representative role for the local people and regarded itself as the authority on their attitudes and reactions. This enabled it to provide an opposition to the European settlers before the rise of African nationalism; so that Kenya was in a sense a two-party state with the Administration, representing the Africans, as one party and the settlers as the other. The political primacy of the Administration in African areas was institutionalized most clearly in local government. D.C.s were chairmen of Local Native Councils (which became African District Councils [A.D.C.s] in 1951) and ran many of their affairs, preparing their estimates, supervising their work and using their meetings as a forum for statements of government policy and for sounding out local opinion.[2]

The third sense of ownership concerned governmental roles. It was not just that administrative officers were considered senior to their counterparts in technical departments. The Administration went further in maintaining a sense of residual right to co-ordinate, oversee or perform any new activity which became necessary. As government became more specialized and more technical officers appeared in the field the Administration sought to control and co-ordinate them in several ways. Most cards played by technical officers could be trumped by an administrative officer by an appeal upwards in the hierarchy. Through his control of Chiefs, Headmen and Tribal Police, and his virtual control of local government finance, he could also largely determine whether any policy or project put forward or undertaken by a technical officer would receive support from the people, or local resources in finance and staff. The most conspicuous response to the growing specialization of government was, however, the formation of co-ordinating bodies. The first of these were the Provincial and District Teams which became prominent during the 1950s and which brought together all local

---

[1] The word 'constituency', now and later, is used in Selznick's sense to mean 'a group, formally outside a given organization, to which the latter (or an element within it) has a special commitment' and which leads to mutual dependence. Philip Selznick, *TVA and the Grass Roots*, University of California Press, Berkeley, 1949, p. 145.

[2] M. N. Evans, 'Local Government in the African Areas of Kenya', *Journal of African Administration*, Vol. 7, No. 3, July 1955, pp. 123–127.

heads of departments under the chairmanship of the P.C.s and D.C.s respectively. Then more specialized committees were formed. Provinces, districts, and sometimes divisions came to have their own Intelligence Committees, Security Committees and Agricultural Committees.[1] Provincial and District Education Boards were set up. *Ad hoc* bodies were formed for special programmes or problems. The pattern was that the administrative officer took the chair and the technical officer (the Special Branch Officer, the Police Officer, the Agricultural Officer, the Education Officer, etc., as appropriate) was the secretary.

The Department of Agriculture ran closely parallel with the Administration. Its hierarchy was fitted to the same land units, with Provincial Agricultural Officers (P.A.O.s), in provinces, D.A.O.s in districts, and often Assistant Agricultural Officers (A.A.O.s) in divisions. Many of its concerns, such as grazing control, soil conservation, and land tenure and use, had political implications with which the Administration was involved. Agricultural officers were dependent on the goodwill of the Administration for the passage of local government legislation to cover agricultural matters, for local government finance, and for support for their extension programmes and special projects. Without assistance from the Administration it was difficult or impossible for an agricultural officer to carry out his functions. There were, in fact, considerable overlaps of roles between the two departments. The Agricolas assumed a political role in assessing for themselves what agricultural changes could be brought about with the local people, and the Administration concerned itself with agricultural matters as being vital to the welfare of the people.

ALDEV[2] was an anomalous department, less professional than the Department of Agriculture but with some similar functions. It was set up in 1945 partly as a reflection of a sense that a new departure was needed in agricultural policies. The ALDEV Board was responsible for financial control of £3 million allocated for the

---

[1] When District Agricultural Committees were set up in 1955 chairmanship was vested in the D.C. by law in Section 138,a) of *Ordinance No. 8 of 1955*.

[2] 'ALDEV' is used here to describe the organization which was successively known as:

1945–46 African Settlement Board
1946–47 African Settlement and Land Utilization Board
1947–53 African Land Utilization and Settlement Board
1953–57 African Land Development Board
1957–60 Land Development Board (Non-Scheduled Areas)
1960–   Board of Agriculture (Non-Scheduled Areas)

African Land Development in Kenya 1946–1962, op. cit., p. 3.

'Reconditioning of African Areas and African Settlement' under the Kenya Ten Year Development Programme 1946-1955. It also fulfilled a co-ordinating function since it brought together the Chief Native Commissioner, the Director of Agriculture and the Director of Veterinary Services. The Executive Officer was an administrative officer, normally a Senior District Commissioner. At first the emphasis was on new settlement but this shifted over the years to a variety of development schemes frequently concerned with water supply and land management. It was a centrally-based organization without a uniform network of offices at the various levels of government in the field, and flexible in recruiting staff and posting them to different areas. The 1950s were a great era for schemes in Kenya and ALDEV, with its funds and staff, provided a means for implementing many of them. Indeed, a foundation member of the Board suggested that the ALDEV crest should be the ALDEV cow being milked by the district authorities.[1] ALDEV was, however, sometimes attacked as ' "the fifth wheel of the coach", an expensive and unnecessary department which duplicates the functions of the Agricultural and Veterinary Departments and the Ministry of Works'.[2] In order to survive it needed strong allies, and managed to prosper through a symbiotic relationship with the Administration, ALDEV staff in districts working for D.C.s. ALDEV's very lack of qualified professionalism made it acceptable to administrative officers who had a 'fellow amateur' feeling, and who were less likely to be trumped by an ALDEV officer using a technical argument than by a qualified agriculturalist. D.C.s made it possible for ALDEV to operate, and ALDEV made it possible for D.C.s to implement pet schemes during the 1950s at a time when many administrative officers were taking an active personal interest in the problems of development.[3]

The description of principal departments is completed by the Ministry of Works (M.O.W.),[4] and especially the Hydraulic Department under the Chief Hydraulic Engineer. Like the Agricolas, the Engineers of the M.O.W. were professionally qualified; but a major difference was that the administrative units of the M.O.W. hierarchy were based on convenience of communication, mainly by road, and often did not coincide with the provinces and districts of the Admini-

---

[1] Ibid., p. 4.
[2] Ibid., p. 4.
[3] An example of the enthusiasm and vigour that could be shown is to be found in D. J. Penwill, 'Paper—The Other Side', *Journal of African Administration*, Vol. 6, No. 3, July 1954, pp. 115–123, describing development efforts in Machakos District, Kenya.
[4] To avoid confusion this department is referred to as the M.O.W. throughout, although at first it was known as the P.W.D. (Public Works Department).

stration. The M.O.W. was more detached from local affairs than the Department of Agriculture, more Nairobi-centred, and more independent of the Administration. It was also often looked upon as something of a poor relation in government.

A further feature of these organizations is less tangible, perhaps less important, but nevertheless of intermittent significance. Different departments attracted and recruited different types of people who brought with them to their departments and jobs a variety of characteristics and attitudes. The Administration was not all public school and Oxbridge; nor were the Department of Agriculture and the M.O.W. entirely staffed with grammar school and red-brick graduates. But there were broad correlations along these lines which could overlay and strengthen interdepartmental tensions, and there was a tendency towards social exclusiveness on the part of the Administration.[1] This was probably less in isolated European communities typical of normal field administration than in larger administrative centres, particularly Nairobi. Tanner has claimed that there was a strong feeling of equality between the small European community's members and, at the time of the civil service's post-war expansion, a real egalitarianism and classlessness.[2] This may have been true of very small communities where people were brought together in social intercourse who would, in their home environments, have found little in common, but was less true where there was a wider choice of association. Typically, departmental tensions were greater the larger the community, and thus the higher in the hierarchy.

Staff for government departments was not found exclusively in the United Kingdom. ALDEV in particular was eclectic in its recruitment, acquiring staff who brought to their tasks a variety of competences and inclinations, at their best combining an inspired practical flair with great enthusiasm and wide experience, if not formal qualification, but sometimes demonstrating the short-comings of amateurs without their virtues. Kenya in the 1950s provided employment for a variety of Europeans who came from many countries to act as prison officers, district assistants, and others, and who ranged from the poor white sadist fascinated by situations of dominance over Africans to the talented colonial wanderer who could and would turn his hand to almost any task.

---

[1] This social exclusiveness can be exaggerated but it is interesting that when on a field visit to Mwea the Governor, Sir Evelyn Baring, called to the Agricolas to join himself and the administrative officers in eating their sandwiches, this was regarded by the Agricolas as a significant and symbolic act of acceptance.
[2] R. E. S. Tanner, 'Conflict within Small European Communities in Tanganyika' *East African Institute of Social Research Conference Papers, July 1962*, p. 4.

Mwea was to have its share of both. Indeed, the most marked feature of the officials who became involved in the Mwea Scheme was their diversity of character.

The idea of irrigation

Like many ideas for development, irrigation on the Mwea emerged and became practicable as a result of a combination of problems, personal and organizational commitments, and a political opportunity. In Embu District in the 1930s there was much discussion of furrows for drawing water from rivers and distributing it for human and animal consumption, and some furrows were constructed and maintained above the Embu-Sagana road. But at that time, when the Agricolas were deeply concerned with conservation policies, irrigation was not a proposition. In the 1940s, as they began to adopt a more positive outlook, the Mwea was seen more in terms of its potential. Apparently the idea of irrigation was first put forward in 1945 when an agricultural officer with experience of rice irrigation in Nigeria proposed a furrow for the cultivation of paddy rice.[1] In view of the post-war metropolitan interest in increasing primary production in the colonies the idea of irrigated rice on the Mwea might well have been seized upon. But the Colonial Office East African Rice Mission which visited Kenya in 1948 reported discouragingly on the prospects for rice irrigation in the Upper Tana basin, which included the Mwea.[2] However in 1949 a Mwea Development and Reclamation Scheme was put forward. Its somewhat conservative title was echoed in the intention, which was primarily to end over-grazing by providing evenly distributed water supplies, introducing controlled grazing, and limiting stock to the carrying capacity of the land; 15,000 acres of eroded land along the Tana were to be reclaimed and a furrow was to be run down the centre of the upper Mwea. The possibilities of the furrow were recognized to include irrigation, but this was to be limited to small areas for experiments, particularly with tobacco and rice. The crying need was considered to be for stock watering facilities to enable the land to be properly used and no large-scale irrigation was contemplated for the time being.[3]

The proposals for the Scheme were accepted and funds were made available through ALDEV. The main commitment to the project was, however, on the part of the Agricolas. In 1949, after

[1] 'The Mwea', op. cit.

[2] Gerald Lacey and Robert Watson, *Report on Rice Production in the East and Central African Territories*, H.M.S.O., London, 1949, p. 28.

[3] Letter, Commissioner for African Land Utilization and Settlement, to all members of the A.L.U.S. Board, 3 January 1949.

consulting some local elders, the D.A.O. established a small irrigation plot in the upper Mwea on the black soils near the Nguka swamp, and for the first time the idea of an irrigation scheme was expressed in a physical form as alterations to land and the growing of crops.

At this time the Administration was deeply preoccupied with political problems, many of which concerned land. This was a period in Kenya 'when politicians and officials were struggling for ascendancy in the Kikuyu country'.[1] The legitimacy of the Administration was felt to rest in its capacity to control, influence and retain the support of the local people. To use force or compulsion was regarded as a last resort and a partial admission of failure. In the atmosphere of deep suspicion of European government in Kikuyu areas in the late 1940s and early 1950s, the Administration was struggling to keep the support of its constituency through combining persuasion and concession on the one hand with threats of compulsion on the other in an attempt to remain plausibly in control.

While the Mwea Development and Reclamation Scheme was being launched the D.C. was trying to secure the Local Native Council's agreement to rules to govern cultivation in the lower Mwea. To this end a body of elders known as the Mwea Committee was formed, but on three occasions it refused to accept any form of rules to control cultivation. The rules were hotly debated in the Council and the D.C. threatened to obtain, and then did obtain, compulsory powers[2] before a negotiated compromise was reached. The position that emerged was that above a certain line (subsequently known as the Wainwright line[3]) land would be divided between clans and individual land ownership would be permissible, while below the line land was to be communally owned and jointly administered by three Chiefs.

At this point it is possible to see that inter-departmental relations had territorial and political dimensions. The D.A.O.'s base was primarily territorial, the D.C.'s primarily political. The D.A.O. was concerned with the agricultural development of the Mwea. He could fairly claim to know the Mwea and its people better than any other district-level government officer. Indeed he intermittently occupied it by camping near his experimental plot. But the plot, however small, constituted land alteration and cultivation carried out

---

[1] M. P. K. Sorrenson, 'Counter Revolution to Mau Mau: Land Consolidation in Kikuyuland, 1952–1960', *East African Intsitute of Social Research Conference Papers*, June 1963, p. 5.

[2] In Government Notice No. 259 of 1950. Quoted in *African Land Development in Kenya 1946–1955*, Nairobi, p. 97.

[3] The Wainwright line corresponded roughly with the alignment (1967) of the Wamumu-Thiba Camp-Wanguru road.

by a European in an area where the Ndia had earlier had difficulty in repulsing at least three European bids to settle. The Mwea Committee was consequently very hostile to the experiments. This drew in the Administration, concerned with maintaining some support from its constituency in the Mwea Committee. With the demarcation of the Wainwright line, negotiated with the Committee, the D.A.O. was forced to abandon his plot since it fell within the area of clan and individual ownership. The Administration and its political constituency thus displaced the Agricultural Department from its territorial base.

Meanwhile, as part of the Mwea Development and Reclamation Scheme, a furrow was being run down the centre of the upper Mwea. It is typical of the twists and turns of development that this furrow defeated its original objective and achieved one for which it was not explicitly designed. The purpose of permitting controlled grazing and cultivation was turned on its head by the increase of settlement which it attracted. A report in 1950 put it that:

> The sight of water proved so tempting to the inhabitants who had their villages in the foothills some miles to the northwest that they immediately started to stake claims in the area being opened up, and to plough up the land with a view to getting a quick return from maize. Considerable inroads were thus made into the grazing area . . . and the Administration have been faced with a difficult problem of sorting things out.[1]

But the furrow went as far as the Wainwright line and the D.A.O. was able to use it to resume his irrigation experiments on a plot on the communal land about a hundred yards below the line. Ironically the furrow made more difficult the control of cultivation and grazing for which it was ostensibly intended, but enabled the idea of irrigation to remain in play.

In these circumstances the Agricolas deepened their commitment to the idea of irrigation. Before being dislodged from his first experimental plot the D.A.O. had shown that rice would grow, and the Agricolas now instituted further investigations. In 1951 ALDEV was called in to carry out a survey of irrigation potential, following which it was thought that some 3,000 to 6,000 acres might be irrigated. Preliminary soil and topographical surveys were carried out, negotiations opened up with the Water Resources Authority for the extraction of water from the rivers, and tentative enquiries made about land acquisition. In the prevailing political climate it cannot have appeared that it would be easy to obtain land, but the idea of an

---

[1] *African Land Utilization and Settlement Report 1945–1950* (mimeo), p. 15. The furrow was dug in 1950.

irrigation scheme had caught hold. At this stage, however, the Agricolas could promote it only as a long-term possibility, not as an immediate proposal.

This position was transformed by polical change. The Emergency declared on 21 October 1952 to combat the Mau Mau rebellion provided both a catalyst and an opportunity. Faced with widespread Kikuyu revolt the Government responded with urgent and decisive action. Large numbers of Kikuyu were 'repatriated' to the Kikuyu Land Unit from the European Settled Area. Many of them were landless and had to be accommodated in camps. Further, the political threat to the European lands which those in revolt sought to acquire had to be diverted. Perhaps remembering the Land Commission's proposals, officials again turned towards the Mwea. Could settlement on an irrigation scheme absorb some of the repatriated Kikuyu and help to contain the Kikuyu population explosion?[1] Could the landless labourers displaced from the European lands be secured, stabilized and controlled on the Mwea? Irrigation may have appeared suitable because of its overtones of discipline and the opportunities irrigation works would provide for the use of labour. The Mwea was appropriate because it was removed from the main population centres but still within the Kikuyu Land Unit. Politically the idea was attractive. The once-empty area that had been so easily moved from district to district when there was little pressure for its occupation now, by virtue of its new potential, assumed far greater importance. In October 1953 the Secretary for Agriculture recommended the building of two camps in the Mwea, and in December the position was that the Scheme was to go ahead as quickly as possible. The immediate purpose was social and political, not economic. It was, through irrigation, 'the resettlement of Kikuyu families *within their own land unit*'.[2] The ideas of conservation and control earlier applied to land were now to be applied to people.

[1] For other proposals and a more detailed description of the position see M. P. K. Sorrenson, *Land Reform in the Kikuyu Country, a Study in Government Policy*, Oxford University Press, Nairobi, 1967, pp. 220–222.
[2] *Department of Agriculture Annual Report 1953*, p. 53. My italics.

CHAPTER 4

Changing the Land

At the end of 1953 the Mwea Scheme existed mainly as an idea, as limited survey information, and as growing departmental and individual commitments. As the Scheme grew the organizations which came together to create it were at first primarily concerned with activities related to changing the land into an irrigation system, and achievement was measured in terms of the amount of construction and land alteration that was accomplished. Although the two departments most closely involved were the M.O.W. and ALDEV, the Administration was influential in pressing for speed on political grounds and in negotiating with the local people for the land. This chapter, in considering organization, activities and relationships connected with the development of the land, is primarily concerned with the years 1954, 1955, and 1956, since it was during those years, before the process of settlement had got under way, that alteration to the land was the centre of attention; but some aspects of land development are considered through to mid-1960 when, for a time, the work of changing the land ceased.

Organizations, activities and achievements

There is no point at which precisely it can be said that a decision was taken to implement the Scheme. Rather there was a rapid flow of events towards involvement and commitment. In December 1953 the ALDEV Executive Committee, in deciding to press ahead with the Scheme as quickly as possible, may well have been doing no more than setting its seal on a matter that had already been tacitly agreed. Already in December a surveyor was on the ground carrying out reconnaissance for siting canals. At this stage preparations for the Scheme were driven forward at great speed, not so much because of the long-term objective of resettling Kikuyu as because of the immediate need to find work for the detainees who were expected to move into the camps early in 1954.

Given the political history of the Mwea and the lack of technical information it was only the extreme conditions of the Emergency which made implementation of the Scheme possible. No effective objections could be raised by the Ndia or Gichugu, although the land to be occupied was in the upper Mwea, much of which was being cultivated by local people and *ahoi*.[1] The upper Mwea, like the rest of Central Province, was pervaded with an atmosphere of fighting and fear. In 1953 there was a battle with Mau Mau on the Mwea, and the Nguka swamp, high with papyrus, was used as a Mau Mau hideout. With detention, curfews, restrictions on movement, compulsory labour and communal punishments among the repertoire of rules and sanctions applicable by government, criticism of official policies was muted. Similarly, at this stage technical objections to rapid development were not accorded any prominence, although the information available was sketchy. Apart from the limited results of the experiments on black soils near the Nguka swamp, almost nothing was known about what crops would grow; and new information was lost when Mau Mau damaged the experimental irrigation plot below the Wainwright line, killed the agricultural instructor in charge, and in burning down his house destroyed the experimental records. Knowledge of flows in the Thiba and Nyamindi rivers was also inadequate, figures being available for only three years.[2] Nor were there any rainfall, humidity or temperature figures for the Mwea.[3] Moreover, maps and previous soil and topographical surveys were of only very limited use. The environment into which the Scheme was to be launched was thus scarcely promising: at the human level it was latently hostile to the occupation of the land; at the physical level it was largely unknown. Neither factor augured well for the future.

Other uncertainties concerned organization. At first there was no organization for the control of irrigation schemes, nor was there in Kenya any previous experience of organization for large-scale irrigation on which to draw. It was not even clear which departments were to carry out which functions. The Secretary for Agriculture wrote that survey was to be arranged by 'ALDEV and/or Hydraulic Engineer (Ministry of Works)'.[4] In such undefined and fluid conditions the Administration was looked to for co-ordination and the Executive Officer, ALDEV, was asked to engage staff as required by the D.C., Embu. This reliance on the Administration was not remarkable, particularly since the Emergency had enhanced its position in Central

[1] Survey of Kenya, aerial photographs of the Mwea, April 1954 series, show considerable cultivation.
[2] F. A. Brown, 'Review of the Mweia [sic]/Tebere Irrigation Project' (mimeo), 3 January 1955, pp. 7–8.
[3] The Irrigation Adviser's report on his visit to Mwea and Tebere Irrigation Projects (typescript), 21 April 1954, p. 3.
[4] Letter, Secretary for Agriculture and Natural Resources to P.C. Central Province, 21 December 1953.

Province at a time when it might otherwise have been weakened by departmental specialization and African political activity. To deal with a colonial civil war a clear chain of command was required through an organization with both security and political competences. The Administration provided this by expanding its overall responsibility for law and order and assuming forceful control of new activities: at first the organization of Home Guards, the screening and rehabilitation of Mau Mau suspects, and villagization; and later, land consolidation and community development. The staff of the Administration, both expatriate and African, was greatly increased and its dominance was reflected in the existence under its control of counterparts to some existing government departments.[1] The Administration acted as though it was in command of almost all government activities including irrigation.

On the Mwea 1954 was a year of confused and vigorous activity. The period appeared to the D.C. as one of 'tremendous ad hoccery',[2] and to a detainee who gained a position of responsibility as a time when 'nobody had a steady mind'.[3] Two camps were set up, one at 'Mwea'[4] and the other at 'Tebere',[5] sited to be conveniently close to the main canals that were to be constructed to take water from the Thiba and Nyamindi rivers. Rough surveys were carried out for the canals and the work of digging them was well advanced by April. In the same month many Kikuyu in Nairobi were rounded up and sent temporarily to camps elsewhere in Kenya, but with the intention that they would later be accommodated on the Mwea. Then in July preliminary surveys were known to have suggested that a very much larger area might be irrigated than had earlier been supposed; 40,000 acres was mentioned and it was thought possible that 10,000 to 12,000 Kikuyu families might ultimately be settled. In July Mwea-Tebere was given priority for survey and planning over the six other irrigation schemes which were being considered or implemented. Unfortunately, surveys and planning could not keep up with the

---

[1] *Main counterparts under the control or strong influence of the Administration*

Main counterparts under the control or strong influence of the Administration	*Equivalent government departments*
Tribal Police and Home Guards	Kenya Police
Rehabilitation	Prisons Department
Community Development	Various, especially Health Department
ALDEV	Department of Agriculture and M.O.W.
Land Consolidation	Various, especially Department of Agriculture.

[2] Interview (1965).
[3] Interview (1966).
[4] At Njega's Dispensary near the Thiba River. See map on p. 73.
[5] At Kimbimbi, the subsequent site of the Scheme Headquarters.

demand for work for the detainee labour, and in the same month excavation stopped on the canals, the alignments of which were realized to be of uncertain value. By the end of 1954, the Scheme had come to consist of maps, plans and survey information, and a growing physical entity on the ground in the form of two detention camps, two large canals in the course of construction, and 10 acres of cultivation and water-use trials.[1]

In these early days decisions had to be taken about the location and form of physical development. A dilemma about which area (Nguka[2] or Tebere) to develop first was resolved by a long distance telephone call from the Director of Agriculture to a former D.A.O. This established that there were more red soils on the Tebere side. Although the original ALDEV survey had been on the Nguka side, Tebere was chosen as the first area in the interests of flexibility since it was not known whether the red soils or the black soils would prove more suitable for irrigation, and if one proved unsuitable then the other could be used. A second decision, apparently taken for reasons of convenience for levelling and ploughing, was to develop the black soils in one-acre fields. Both these decisions, being concerned with virtually irreversible physical works, were to determine much of the form and organization of the Scheme in its later history.

Formal central organization came about after, not before, activity. In Nairobi co-ordination was initially carried out at informal meetings about once a fortnight in the office of the Director of Agriculture. In June 1954, however, he considered that a body was needed to plan the various irrigation schemes that were in hand and more particularly to co-ordinate their hydraulic and agricultural aspects. The Joint Irrigation Committee (J.I.C.) which resulted was chaired by the Director of Agriculture, with the Executive Officer, ALDEV, representing ALDEV, and the Hydraulic Engineer, M.O.W., representing his department, as members. The J.I.C. had no direct statutory powers but from June 1954, when the first meeting took place, functioned as the central forum for deciding general policies and for resolving problems between departments. Meetings were held regularly, usually at monthly intervals, and attendance expanded to incorporate a variety of interested parties, sometimes including departmental officers from the Scheme.

At first there was no formal organization for co-ordination at the local level. The Scheme was serviced and created by several departments and individuals with bases that were physically distant, and for which Mwea was only one of a number of commitments. The

---

[1] F. A. Brown, op. cit., p. 3.

[2] This side was at first known as 'Mwea'. This confusing usage was later changed to Nguka, which will be used here throughout.

activities of co-ordination, survey, planning, accounting and labour control were all performed by outside organizations. Co-ordination was effected by the J.I.C. in Nairobi, by a peripatetic irrigation adviser, and at the local level by the D.C., if at all. Topographical survey was the responsibility of the M.O.W. until the end of 1954 when its team was suddenly withdrawn and sent to the lower Tana to explore the possibilities of even larger-scale irrigation. The Survey of Kenya, another Nairobi-based organization, was responsible for the survey control. Soil surveys were carried out by the Government Soil Chemist, who was also from Nairobi. Planning was the work of the Hydraulic Branch of the M.O.W. and was initially done in its drawing offices, again in Nairobi. Accounting was undertaken by the staff of the D.C.'s office in Embu, though also by a Prisons accountant on the Scheme. Even the camps were no more than an extension of the activities of the Prisons Department. It was only in the presence of a few ALDEV staff, described as Works Supervisors, that at the end of 1954 there existed the germ of a separate scheme organization.

The need for more effective local co-ordination was recognized, but ideas about how it should be achieved varied. In late 1954 representatives of the Administration and the Prisons Department considered setting up a Mwea-Tebere Development Board to be chaired by the D.C. An adviser called in from Gezira at that time made a different suggestion. He observed that there appeared to be little co-ordination of effort between the various departments and recommended that 'some person with authority should at this stage be made responsible for development as a whole'.[1] An immediate solution was sought, however, not in the appointment of an overall commander of operations but in the formation of a committee. In January 1955 a meeting of the ALDEV Board with a number of administrative officers, agriculturalists and hydraulic engineers, resolved to set up a local committee, chaired by the D.C., which would issue instructions to the various technical officers on the Scheme; these would, however, have direct access to their heads of department and to the J.I.C. on all technical matters. The D.C. wrote after the meeting that he intended to take a personal interest in the Scheme's development, and it appeared that he would be the principal co-ordinator.

The Director of Agriculture was not entirely satisfied with this arrangement. He felt that even with the best will in the world, the D.C., with his many other commitments, would not be able to devote the necessary time to the project. Indeed, although the D.C. was spending about a quarter of his time on the Mwea his main

[1] F. A. Brown, op. cit., p. 21.

preoccupation was with the camps and rehabilitation. In January 1955 the Director of Agriculture argued the ultimate need for an irrigation manager attached to the Department of Agriculture but responsible to the committee, to take charge of the Scheme and handle day-to-day matters. In March he went further and favoured the immediate appointment of an overall manager to effect co-ordination between the various departments involved. He wanted someone on the spot who would keep him in touch with developments and see that as far as possible nothing contrary to good agricultural principles was done. No objection was raised by the Administration, and at the end of March the peripatetic irrigation adviser was posted in to the Scheme as the first Manager. He was paid by ALDEV, but in practice worked to the D.C. at the local level and to the Director of Agriculture in Nairobi.

With the arrival and early activities of the first Manager the emergence of the Scheme as an entity was taken a step further. The three ALDEV Works Supervisors already in the area now worked under him, forming a nucleus of staff which was soon expanded. In May 1955 another Supervisor was recruited and in November an Assistant Manager joined the Scheme. The Scheme was becoming an organization in its own right. Meanwhile the Local Committee[1] began to function formally, recording its first meeting only on 25 August. The Scheme's resources grew as five more camps were constructed (1955) and more detainee labour became available. Most significantly, the Manager introduced a new dimension soon after his arrival by starting to prepare land for irrigation. The Nguka swamp was drained, partly to reduce malaria, and the papyrus which grew on it was burnt because it provided a hide-out for Mau Mau. The Manager then began to lay out one-acre fields and to grow rice. In the conditions of haste, confusion and inexperience which prevailed, physical achievements fell far short of targets and expenditure was high. By mid-1956 recorded expenditure already totalled £173,000,[2] but by the end of the year the achievement in land preparation was only 589 acres.[3] However the headworks for the Tebere Section had been completed by an outside contractor, water use experiments and crop trials were in progress, and the Scheme had begun to exist as a seasonal agricultural process through the cultivation of paddy rice, which in 1956 was sold for a

[1] Known as the Mwea-Tebere Irrigation Scheme Committee until 1960, and then renamed the Mwea Irrigation Settlement Committee, but referred to here throughout as the Local Committee.
[2] 'National Irrigation Schemes: Consolidated Capital, Expenditure, and Revenue Figures for the 11 years to 30 June 1966 (amended)', op. cit. The exact figure was £173,024. This included expenditure on works camps and water supplies.
[3] Annual Report Mwea/Tebere Irrigation Scheme 1957.

revenue of £7,000.[1] The Scheme had come to be not just information, surveys, and plans, but also altered land, a small organization, and a farm in production.

The physical development of the Scheme depended not only upon the activities which were carried out on the land, but also of course upon the resources made available. Of these finance was antecedent to others since it was required to pay for machinery, tools, staff and other items essential to the work that had to be done. Without any prospect of substantial revenue in its early years the Scheme was dependent on continued financial support not only for expansion but for its very survival. At first there was no shortage of funds. During the crisis years of the Emergency there was more money available than could be used. At the same time it was too early in the life of the Scheme for economic assessments to be made. But from 1956 onwards there was mounting official concern about expenditure. In part, this may have been because the Scheme was becoming more identifiable as a separate physical, economic and organizational entity which could be considered as an independent unit to be assessed. This concern certainly reflected a more general tightening of accounting controls following the laxity of the height of the Emergency. In part, too, concern arose from the high expenditure for small statistical achievement.[2] The 1954 Swynnerton Plan Proposals for the Scheme had been for £150,000 capital expenditure spread over five years;[3] but actual expenditure on the Scheme by mid-1957 was £364,000. By the end of 1957 the estimates for the financial year 1957/58 had been cut by about one-third (from £326,860 to £216,474), but current expenditure must still have appeared heavy in relation to the achievement that could be presented in the form of statistics. At the end of 1957 the total acreage developed was only 846[4] (against an official estimate given a year earlier of 4,000 acres by this date[5]), no settler had been given a permanent holding (though 167 were on trial two-acre holdings), and the Manager observed drily that the year's work could not be considered an unqualified success.[6] At the same time, although rice on the black soils was showing promise no system of water application had been agreed for the red soils and no suitable crops had been found for them. The political importance of the Scheme was such that there was never any question of closing it down. But as economic

[1] *Department of Agriculture Annual Report 1956*, pp. 64–65.
[2] See Table 4.1 on p. 66.
[3] *A Plan to Intensify the Development of African Agriculture in Kenya*, compiled by R. J. M. Swynnerton, Government Printer, Nairobi, 1954, p. 39.
[4] Annual Report Mwea/Tebere Irrigation Scheme 1957, p. 2.
[5] *Department of Agriculture Annual Report for 1956*, p. 65
[6] Annual Report Mwea/Tebere Irrigation Scheme 1957, p. 6.

criteria began for the first time to be applied a crisis developed over the size to which it should be expanded.

TABLE 4.1

Mwea-Tebere Irrigation Scheme: total recorded expenditure against reported achievements in prepared land

Year	Total capital and recurrent expenditure recorded to 30 June £	Acreage fully prepared, totals reported by the end of the calendar year		
		Red soils	Black soils	Total
1955	n.a.	n.a.	300	c.300
1956	173,024	26½	562½	589
1957	364,281	171	675	846
1958	550,163	252	1,749½	2,001½
1959	694,741	252	3,634	3,886
1960	1,004,951	252	4,974	5,226

Sources: Expenditure figures are those given in 'National Irrigation Schemes: Consolidated Capital, Expenditure, and Revenue Figures for the 11 Years to 30 June 1966 (amended)', op. cit.

Acreages are those given in Mwea/Tebere Irrigation Scheme Annual Reports for 1957, 1958 and 1959, and Mwea Irrigation Settlement Annual Report for 1960, except for the 1955 figures which are derived from *African Land Development in Kenya 1946–55*, op. cit., p. 104.

Official reactions varied. One response was protective. Those who were concerned with the future of the Scheme as a viable entity were anxious to limit that proportion of expenditure which might be charged to it as a loan. The J.I.C. resolved that the expenditure charged in the early stages of the Scheme to a grant from the United Kingdom and to funds from the United States aid agency should be treated as a free capital grant without interest chargeable to the Scheme, and that the poor performance of detainee labour should be taken into account in assessing the relative sizes of the grant and loan elements in the capital expenditure attributable to the Scheme. A second response was to cut estimates and, with them, the size to which the Scheme might be developed. In early 1958 the Treasury was exerting pressure for a 40% cut in the estimates for 1958/59, and to meet this cut targets for development were reduced. But this stringency had other, secondary effects in forcing rationalization on the Scheme. The expensive and controversial development of the red soils was abandoned, to be followed by a decision to start work on the black soils of the Nguka side in addition to Tebere. The system of labour gangs was given up for piece-work, both on field development and housing construction, leading to a marked improvement in productivity. But one proposal, to reduce the labour force, was vetoed by the Governor who took a close personal interest in the Scheme. Instead he sought from the United Kingdom Government the £250,000 which was considered necessary for the completion of 5,000 acres. When this sum became available in 1959 there followed a period of hectic activity in which this area was completed by the end of the 1959/60 financial year.

The great differences between expenditure and recorded achievement between the period up to the end of 1957 and the period from 1958 onwards can be explained in terms of many factors. Some were peculiar to Mwea. The detainee labour force dropped from a peak of nearly 7,000 in 1956 to about 2,500 by the end of 1957, and became insignificant in 1958.[1] This form of labour was said to be very inefficient. The unpredictability of the phasing-out of detainee labour led to difficulties in transferring to paid labour and to machines. Staff morale, also a variable, was improved by the 1957 short rains rice crop on Tebere which was successful enough to suggest that the Scheme would have a feasible economic basis.

But other factors were more general, and apply to any scheme requiring heavy capital investment. First, many capital works had to be undertaken which did not appear in statistics: the construction of roads, bridges, staff housing, water supplies, headworks, feeders and drains did not appear in tables—some, such as staff housing, because they were individual items and some, such as feeders and drains, because they were not easy to measure. Second, the time taken to complete a major work meant that expenditure came to book before the work appeared as a completed item. Yet other factors were directly associated with the pressures for speed, the administrative confusion, and the financial permissiveness which existed in Mwea as in many other settlement schemes. Thus, a third factor was poor accounting and lax controls over expenditure. Fourth, the lack of technical information and expertise allowed the pursuit of unprofitable lines and over-investment in development that was really only experimental. Fifth, inefficient methods were permitted to survive. In these ways, as commonly with other high capital projects which are forced through at great speed and without careful planning, the costs were considerably higher than they would have been with more deliberate and systematic investigation and development. This is, however, to state a cause (speed) and an

[1] *Department of Agriculture Annual Reports*, 1956, 1957 and 1958.

effect (higher costs) somewhat baldly, without discussing the intervening processes. It will be useful, therefore, now to examine in more detail the nature of the pressure for speed and some of its more immediate effects.

The pressure for speed: political versus technical considerations

Many of the early problems of the Scheme can be traced to the urgency with which it was pressed forward. This in turn resulted from its political purposes, as seen by the Administration, to give work to the detainees and in the longer run to provide land and employment for the maximum number of landless Kikuyu families. This latter became an increasingly important aim. Land consolidation in Central Province was establishing individual title to land but displacing *ahoi* who became candidates for resettlement. Further, as detainees passed through the Mwea pipeline and were returned to camps in their home districts it became evident that Chiefs and others were not prepared to welcome the release of some men to their home areas, and Mwea was seen as a possible alternative destination for them. Simultaneously, the United Kingdom Government was pressing for faster release of detainees and the rehabilitation pipeline was found to have a more rapid through-put than had been expected. This both limited the period during which detainees' labour could be used on the Scheme, and multiplied the numbers who might have to be resettled. For all these essentially political reasons the Administration continually urged the technical departments to speed up their operations and expand the proposed acreage.

This pressure for speed encountered technical objections. With its amateur approach and dependence on its alliance with the Administration, ALDEV did not put up much resistance to fast development; but the more professional departments, the Department of Agriculture and the M.O.W., were reluctant to agree to rapid expansion. The Agricolas were uncertain of irrigation techniques, crops, and methods and timing of cultivation, and wanted more time to experiment. Seeing themselves as the future managers of the Scheme they were anxious to avoid responsibility for an agriculturally unsound enterprise. For their part, the Engineers were uncertain about the structures they were designing and were anxious for time to observe their performance before designing more. They feared the permanent, indeed monumental, visibility of any mistakes they might make and were able to point to the abortive work carried out on the canals in 1954 as a warning of the results of haste. Nevertheless, despite these technical objections and uncertainties, the Administration was able to use its dominant position and the political rationale of the Scheme to ensure that it was driven forward with great energy.

One effect of this urgency was a confusing overlap of activities which would normally have occurred in sequence. On strictly technical grounds, an irrigation scheme should be created through an ordered series of activities: topographical survey should precede soil survey, for which it provides the maps; soil survey should precede the design of the irrigation layout, which should take soils into account; and design should precede the construction which it plans. In the event, the pressure was such that each of these procedures was ignored at one time or another. In 1954 aerial survey and ground control were carried out simultaneously with soil survey. In 1955 irrigation designs were produced as intelligent guesswork by the staff of the Hydraulic Engineer before the completed soil maps were available. In 1955 irrigation layouts were begun by ALDEV in the Nguka swamp without any planning by the Hydraulic Branch.

A related effect was a series of acute bottlenecks. The pressure for speed was such that no sooner was one bottleneck widened than another appeared. In early 1955 a delay over the soil survey results held up the design of the layout, until the latter went ahead regardless. In April 1955 the main problem was a shortage of labour. By May, with 400 detainees arriving a week, the problem was insufficient staff to give them work: in June, for example, both an M.O.W. Surveyor and an ALDEV Works Supervisor were engaged full time laying out contour surveys for bunds but were unable to keep ahead of the labour groups working on land preparation.

A further serious effect of urgency was to aggravate organizational stress and conflict. Where departments were carrying out operations in a necessary sequence, pressures for rapid results were transmitted down the line, with the Administration urging ALDEV to speed up field development, ALDEV pressing the M.O.W. for designs, the M.O.W. calling for soil maps, and so on. Departments also competed for resources, notably labour and staff, which were vital for the rapid execution of their work. Further, the lack of time for consultation led to activities which should have been co-ordinated proceeding independently.

Faced with the Administration's pressure for speed, one defensive response of the technical departments was to call in advisers from abroad. In advocating slow development experts from outside would, it was felt, carry more weight than local officers who were known not to know much about irrigation. One adviser came from the Gezira Scheme and two from the Colonial Office. All, as expected, emphasized the need for gradualness in development. The first (from Gezira) wrote that: 'The Scheme is promising, but should not be pressed on too quickly. There is little knowledge available as to suitable varieties of crops to be grown, and how either they or the different types of soil will react to irrigation, or how much water will be required or when.'[1]

One of the Colonial Office advisers, an engineer, was more outspoken:

> The Mwea-Tebere project presents exceedingly unusual features inasmuch as the project has been undertaken before topographical surveys have been completed, before water requirements have been estimated, and before a soil survey has been made. Add to this the fact that no headworks have been constructed and that no engineering plans, as I understand them, are available, and my task as an engineer on being asked to comment on the project can be appreciated. This is the only project I have encountered where labour has been imported into the area and attempts made at canal construction before the collection of all essential information without which it is impossible to produce a really satisfactory engineering project.

He went on to warn that: 'The idea that a large number of families can be imported into an area of this kind and that these families will rapidly acquire the technique of irrigation and become entirely self-supporting must be rejected.'[2]

All advisers agreed that on technical grounds the Scheme should be maintained as a small pilot project for several years in order to gain more information and experience.

Political considerations were, however, supreme. From the Governor downwards in the Administration the view was taken that the Mwea was important for its own sake as a productive scheme, but that it was even more important as a means for the reabsorption of the many displaced Kikuyu who presented Government with its most acute and difficult problem. In the event, the technical departments co-operated with the pace of development required on these political grounds. That they did so can be attributed to three influences. In the first place, the Administration dominated policy during the Emergency and could always in a last resort apply *force majeure* through its hold on central government. Second, the political and security argument could be used most effectively by technical officers themselves in order to obtain resources for their work or to expand their staffs. For instance, in 1956 when the Hydraulic Engineer wanted to limit development to a 500-acre pilot project and the Administration was pressing for expansion of the Scheme, agreement on expansion was reached when the Administration undertook to help the Hydrau-

---

[1] F. A. Brown, op. cit., p. 21.
[2] Note by Gerald Lacey, Drainage and Irrigation Adviser, Colonial Office, on the Mwea Tebere Irrigation Scheme, 8 February 1955.

lic Engineer to recruit the extra staff he said he needed. The third factor which persuaded technical officers to act contrary to their professional conventions was the knowledge that they could always exculpate themselves from the consequences of their actions on the grounds that they would never have been undertaken but for political pressures. Speed insisted on by the Administration thus not only provided them with easy access to resources; it also gave them, as it were, an escape clause in case of failure. To this extent the Administration, the Engineers and the Agricolas all had a common interest in maintaining the Scheme's security and political rationale.

Departmental colonization

The process of forming the Scheme can be regarded as a new layer of colonization, as a new pattern of occupation superimposed upon, dislodging, and to some extent incorporating that of the Ndia, Gichugu and *ahoi*. It was now, as before, a frontier situation in which men competed for property and sought to define boundaries. Earlier the Ndia had asserted their rights to land use by knocking Mbere beehives out of trees with stones and by public pleading to the Kenya Land Commission. Faced now with a far stronger threat in the form of government and the Scheme, they were at first unable effectively to protest. Indeed, in the early stages of the Scheme conflict between local people and government was less conspicuous than conflict within government itself. As with tribal groups earlier, so now with organizations, undefined ownership presented an incentive for aggressive entry and occupation. But in this new colonization there were differences. As weapons to displace opponents paper was used in place of stones, and cases were put to committees in place of speeches to a Commission. The most obvious 'property' for which the organizations competed was also different: the tribal groups had appropriated land; the organizations which followed them appropriated roles and the resources needed to perform them. The contrast between the earlier territorial and the later functional colonization could, however, easily be made too sharp. At a less obvious, but nevertheless important, level the behaviour and attitudes of the officials to one another, as of the local people who preceded them, were influenced by their geographical distribution and their relationships with the land.

Throughout the history of the Scheme the spatial dimension had a significant bearing upon events and organizations. It is, however, so basic and so obvious a factor that it is easy to overlook it. Yet in the early days of the Scheme it both reflected and influenced the attitudes of officials to one another. Seven camps were built in the Mwea;[1]

[1] See map on p. 73. The camps were Mwea, Tebere, Kandongu, Thiba, Gathigiriri, Wamumu and Karaba.

each had accommodation for about a thousand inmates and had a European staff of three to five officers who lived in the camps, with their wives when married. For security reasons all houses had to be built within the camp perimeters, but there was no shortage of space for additional housing. Nevertheless, as officers from different departments came to work on the Scheme, they settled in different camps. In 1954 an Agricultural Research Officer was posted to the Scheme and went to live in the 'Mwea' Camp, probably because it was the one of the first two built, although his experiments were some 12 miles away. When the Manager came in March 1955 he lived first at Wamumu and then later at Thiba Camp. When the Resident Engineer of the Hydraulic Branch of the M.O.W. arrived in June 1955, he moved into Gathigiriri Camp. The Assistant Manager, who came in November, established himself at Tebere Camp, which later became the Scheme headquarters. When a D.O. was posted in he lived first at Karaba Camp, then at Thiba, before finally setting up an entirely new centre at Wanguru on a 25-acre site.

At first sight this dispersion appears to reflect a syndrome common to many colonial wanderers. Given that the impulse to lead a colonial type of existence was often misanthropic, avoidance could be expected as a frequent response to situations in which roles and relationships were not clearly defined.[1] People who had travelled and lived in varied and sometimes isolated situations might tend anyway to be self-sufficient. From this angle, those who came to staff the Scheme were perhaps not exceptional. The first Manager had lived in India as a Zamindari and had worked in Eritrea before coming to Kenya. The Assistant Manager also had a long background in India. The D.O. had served for many years in the Sudan and came to the Scheme out of retirement. To these differences of background it may be added that the Resident Engineer was an Irishman, the Agricultural Research Officer an Englishman, and at least one ALDEV Works Supervisor had had experience on Gezira. In some of the recreations of these men—gardening, shooting birds or buck for the pot, fishing, for instance—there was certainly a strong individualist element. Moreover, although money was voted for a club at Tebere it was never built. In a sense the lawn of the Manager's house, levelled to make a tennis court that was never played on, symbolizes social relations that were contemplated but rejected.

[1] Avoidance was expressed in postings policies. Transfers as a means of dealing with conflicts in small European communities in colonial situations are discussed by Tanner (op. cit., p. 3). Avoidance responses to difficulty may well have been more common in colonial civil services than others and may have been responsible for some of the lack of staff continuity that is often levelled as a criticism of colonial personnel practices.

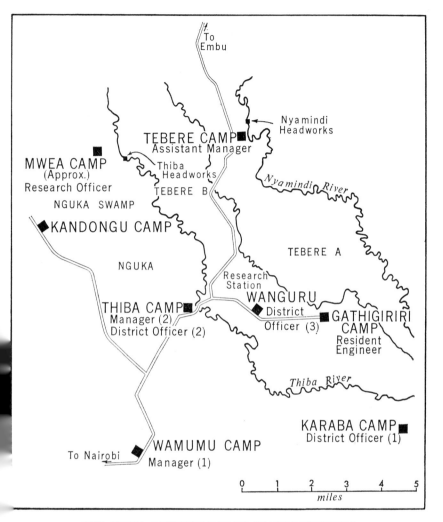

To
Embu

Nyamindi
Headworks

TEBERE CAMP
Assistant Manager

MWEA CAMP
(Approx.)
Research Officer

Thiba
Headworks

Nyamindi River

TEBERE B

NGUKA SWAMP

KANDONGU CAMP

NGUKA

TEBERE A

Research
Station

WANGURU
District
Officer (3)

THIBA CAMP
Manager (2)
District Officer (2)

GATHIGIRIRI
CAMP
Resident
Engineer

Thiba River

KARABA CAMP
District Officer (1)

To Nairobi

WAMUMU CAMP
Manager (1)

0   1   2   3   4   5
miles

THE MWEA SCHEME: EARLY STAFF DISPERSAL

But this is no more than a partial interpretation. There was also some of the conviviality popularly associated with small expatriate communities. At one period there was a Scheme party once a month. At least two tennis courts, one of them on a rice-drying floor, were used. There was an open invitation to play croquet on Sunday mornings on the Assistant Manager's lawn, and his wife organized a wedding reception when one was needed and gave the largest party ever held, for 72 people, to say farewell to a popular rehabilitation

73

officer. All the same the general impression of social life, particularly in the early years of the Scheme, is of greater individual social isolation than might have been expected in a typical small expatriate community.

Geographical and social isolation evidently reinforced each other. Official and unofficial communication was irregular and unpredictable so that co-ordination between officers and departments and the resolution of problems were more difficult than they might otherwise have been. Not only were there few informal opportunities for contact but, until the Local Committee was convened, there was no regular formal contact either. There was no telephone, and it could be easier for an officer to communicate through the Prisons wirelesses with his headquarters in Nairobi than with the staff of another department on the Scheme. For when it was essential for one person to consult another he could spend several hours driving round the Scheme on bad roads looking for him only to find that he had gone to Embu or Nairobi or was himself driving round the Scheme looking for someone else. Problems were saved up for meetings of the Local Committee or J.I.C. at which there was a tendency for departmental postures to be adopted and departmental 'lines' argued.

Territoriality in animals and man is a complex phenomenon, and it would be misleading to attribute interdepartmental behaviour largely to it. Possessiveness over land was only one of several levels on which behaviour and attitudes can be interpreted. Nevertheless, departmental officers on Mwea, as on some other schemes, were carrying out actions which had overtones of land appropriation. In different ways the Agricultural Research Officer, the Manager and his staff, the M.O.W. and the D.O. can all be seen as acquiring territorial rights through their various activities in changing land. The Agricultural Research Officer developed his research station near Wanguru as a separate colony: staff housing, sheds, and experimental fields were set out as a distinct and bounded entity and the land was separately set aside.[1] The Manager lived on the Nguka side, and at an early stage began development of the Nguka swamp over which he then exercised almost complete control. The Resident Engineer concentrated on the Tebere side where he lived, and was able to extend his supervision and control of the process of changing the land to an extent which he found impossible in the Nguka swamp. In setting up his own separate headquarters at Wanguru the D.O. asserted his independence, and at a personal level acquired and developed territory by making an exceptionally beautiful irrigated garden. The invasion of the Mwea by departments thus created a number of new colonies and settlements occupied by different departmental officers,

[1] In *Kenya Gazette Notice 3428*, 26 July 1960.

74

providing them with headquarters close to land in which they acquired special interests and over which they came to exercise special degrees of control.

The spatial separateness of departments, the difficulties of communication which it aggravated and the attitudes which it expressed and encouraged, have to be borne in mind in considering the organizational relationships which evolved. These factors accentuated interdepartmental problems but did not necessarily create them. The origins, growth and nature of these problems were more complex and varied, and were closely linked with the roles and interests generated by the growth of the Scheme.

Engineers and Agricolas: the M.O.W and ALDEV

In order to create the physical form of an irrigation scheme the activities of survey, design and construction have to be carried out. On a scheme such as Mwea, design and construction can be further broken down according to the different components of the irrigation system into headworks, canals, feeders, fields and drains. The process leading up to and including the physical changing of the land can be regarded as a flow of phased activities which require resources for their execution and which bestow roles upon those who perform them.

In the case of Mwea, the two groups most clearly involved in the activities necessary for changing the land were the M.O.W., represented by the Resident Engineer and his staff, and ALDEV, represented by the Manager and his staff. There was sufficient coherence of attitude among the staff of each department for it to be possible without undue distortion to describe departmental attitudes. Since the approaches of the two staffs were associated with professional and functional differences, the M.O.W. will be described as 'the Engineers'; and the ALDEV staff as 'the Management' in matters relating to the running of the Scheme and as 'the Agricolas' in matters agricultural, where their attitudes coincided with those of agriculturalists of the Department of Agriculture.

In the early stages the roles were unoccupied and unowned. For example, it was not clear at first whether the topographical survey would be carried out by the ALDEV Management or the M.O.W. Engineers. In fact, in this undefined situation functions could be performed by whichever department had the necessary resources. There was too much to be done too urgently for overlaps to be serious. The Engineers at first carried out the topographical survey, and the Resident Engineer was fully extended on canal design and then on the design of field layouts for the Tebere section. He was also responsible for supervising the contractor who built the headworks

75

on the Nyamindi river. Although what the Engineers could undertake was limited by shortages of staff they claimed the right to survey and design the whole irrigation system. The Manager, however, was anxious to proceed rapidly with development. Soon after his appointment in March 1955 he ignored the professional advice of the Engineers and prepared the Nguka swamp for cultivation. With the amateur flair of ALDEV he dammed and diverted a stream, drained part of the swamp, levelled fields and began to grow rice. He thus performed the whole sequence of activities—survey, design, layout and cultivation—for that particular area.

For other areas of the Scheme, survey and design were carried out by the Resident Engineer and his staff; the plans were then handed over to the Management to implement. In practice, there was considerable friction over this and over subsequent handover points. The Engineers considered that the Management were unable to follow their plans accurately, and carried out an uneasy form of supervision. The Resident Engineer acted as a sort of consultant to the Management and had some scope for oversight of field preparation and a watching brief over the work carried out by the Management Works Supervisors. But construction proceeded very slowly, and from late 1955 onwards the Manager and his staff were distracted by issues arising from the arrival of settlers and the growing of crops. In these circumstances the Engineers were able to push their control of operations along the sequence of activities and also geographically down the irrigation system. From the start the Engineers had supervised the contractor who built the Nyamindi headworks. By January 1956 the Manager agreed that the Engineers should supervise canal construction. After slow progress with field preparation under Management supervision in 1956, the Engineers' responsibility was extended all the way through to the end of field preparation. The Engineers thus followed their control of the process through from a start with survey and design to include all construction work, with a corresponding shift of the point of handover to the Management from the end of design preparation to the completion of structures and fields.

A continuation of this process would have led to control by the Engineers of water supply and water use. This was suggested by the Hydraulic Engineer in December 1955 and at one time it was thought that a permanent member of the M.O.W. staff on the Scheme would supply water to the Manager. That this never came about may have resulted from the close association of water control with cultivation, which was clearly a Management responsibility, and from the full extension of the Engineers' resources on their expanded activities. In the event, the frontier between the two departments, the M.O.W. and

ALDEV, was stabilized at the handover of completed fields, though even on this boundary there was friction and misunderstanding. Nevertheless the pressures and adjustments of the years 1954 to 1957 produced a relatively stable system for design, construction and handover, so that when the bulk of the development work took place in the years 1958–1960 the activities had become routine, more efficient, and less noticeable from an organizational point of view. As the ownership of roles and their boundaries were established, stress was reduced.

Throughout these changes each department sought to achieve independence of the other. The Manager's early attack on the Nguka swamp can be interpreted as an act of attempted self-sufficiency. For their part the Engineers valued separation, an organizational form of avoidance action. In the latter part of 1955, when problems had arisen over the supervision of canal construction (the Resident Engineer having the expertise needed, and the Manager having the supervisory staff) the Hydraulic Engineer argued that responsibility should be specifically allocated and that 'Insofar as it is possible, the Resident Engineer should be in complete control of his own sphere of activity and not be dependent on the Irrigation Manager more than is necessary in respect of making use of any services which cannot reasonably be split'.[1]

He proposed that the Manager should be responsible for land preparation, crop production, communications and temporary housing, and the Resident Engineer for canal layout and construction, supervision of contractors, control of water and ancillary engineering operations. In practice, however, complete separation was impossible and the two departments were driven into co-operation. When the Resident Engineer was short of staff he borrowed two Works Supervisors from the ALDEV Management. That the arrangement was unsatisfactory does not reduce the interdependence it reflects. Again, it was convenient for M.O.W. tools to be stored by ALDEV, and the M.O.W. made use of the more flexible ALDEV accounting system for quick purchases. Co-operation might also be called forth by a crisis: when a canal bank burst and attempts to seal it with sandbags had failed, an ALDEV and an M.O.W. Supervisor worked together in an ingenious attempt to stop the flow by placing a strong log across the breach and throwing in bamboos at a slant on the upstream side. In fact ALDEV's handyman ingenuity and the M.O.W.s professionalism were complementary. The M.O.W. Engineers' expertise was needed by the ALDEV Management for the supervision of the more complex works and ultimately for the rapid completion of field preparation, but the Management's practical

[1] Letter, Hydraulic Engineer to Director of Agriculture, 13 December 1955.

flair had also a contribution to make. In a sense, the Management had the vision and the Engineers had the rules.

Other issues between the two departments were associated with constructor-user relationships. As designers and constructors the M.O.W. Engineers had a proprietary interest in the work they had carried out. This was institutionalized in their maintenance role: they were especially concerned that what they had created should not be damaged or altered. The ALDEV Management, however, were concerned that the irrigation system should perform its intended function of conveniently carrying water on to land. One common complaint by the Engineers was that the Management damaged canals by running too much water down them. Another sensitive point was alteration of works. In 1958 the Management blocked some channels in order to raise the water high enough to run it into some fields. The Engineers protested, and their case was supported by the J.I.C. which resolved that no major or minor improvisation of water control should take place without prior reference to the Resident Engineer.[1] The Engineers thus remained to some extent the 'owners' of the works. A further problem arose when it was found that some strips of fields were planned so that there was a field of less than standard size left over at the end: this was inconvenient to the Management who stated that it would prevent equal distribution of land to tenant cultivators. Again, in field design the Engineers had an interest in low capital cost and ease of construction, in contrast with the Management's interest in low running costs and ease of operation. The Engineers consequently preferred to place fewer drops and inlets in the feeders while the Management preferred a separate inlet and outlet to each field, requiring more drops to be constructed but making subsequent water control simpler. Again, at the point of handover between departments, it became clear that matters of concern to the user might appear less important to the constructor. For rice irrigation close control of water levels is imperative for high production. On a paddy field a soil level variation of a few inches can make a considerable difference to yields. Some fields were handed over in a condition less than level until it was decided that they would be flooded to test levelness before acceptance by the Management.[2] Perhaps the most potent tension of a type to be found in the relations between those who design and construct and those who use concerned recognition for achievement. The Engineers felt that although they created the irrigation system the credit went to the Management who were patted on the back for growing crops.

[1] J.I.C. minute 64/58 of 1 October 1958.
[2] Particularly from 1958 it was not intended that fields should be completely level at handover, but flooding made any major irregularities obvious.

In the words of one, 'We do the work and they get the gongs'.[1]

Beyond these tensions and differences some difficulties arose primarily from different professional preoccupations and approaches. This can be illustrated from the deep, almost bitter, division of opinion about the layout of the red soils.[2] Although the commitments of individual officials to the alternative systems went beyond any point at which they could be explained purely in terms of technical arguments, the origins of the disagreements were rooted in the opposed professional interests of the Engineers, concerned primarily with hydraulics and water, and of the Agricolas, concerned primarily with agriculture and crops.

The Engineers were preoccupied with water and water use. They argued that as there was more irrigable land in Kenya than there was water to irrigate it, a prime consideration in designing a water application system should be economical water use. The most economic way to apply water was held to be by running it down furrows. In order to ensure even application these furrows should lie on planes, the layout being known as the inclined plane system.

The Agricolas were preoccupied with growing crops and with the management of the tenants who were to grow the crops. They foresaw grave difficulties in handling the relatively complex inclined plane system since it required individual water control for each furrow. The level basin system which they advocated would have the advantage of only one inlet and one outlet per field, and in addition would involve less loss of fertility through the movement of earth in field preparation.

Neither method proved satisfactory. The inclined planes were difficult and expensive to lay out. As for the level basins, the soils were so porous that it was extremely difficult without using excessive quantities of water to flood them so that there was an equal water supply to each part. Compromises were proposed and tried, areas were laid out according to both systems, decision overtook decision. Eventually it was arranged that the heads of departments should visit the Scheme and observe trials of the various methods. On the day the experiments were washed out by heavy rain. The problem was then buried by the formal decision in March 1958 to stop development of the red soils[3] and to concentrate on the development of the black soils for paddy rice production.

Some of the actors involved in these interdepartmental relations at

---

[1] Interview (1965).

[2] These differences were discussed repeatedly in the J.I.C., especially in 1957, and were the subject of numerous letters and memoranda.

[3] Small-scale experiments on the red soils continued and were still in progress in 1967.

scheme level attributed their difficulties to 'personality clashes'. But given the lack of co-ordination, the pressure for quick results, the spatial separation, the uncertainty about roles, the competition for resources, the difficulties of handover, the constructor-user stresses and the conflicts of professional outlooks—all of which tended to provoke interdepartmental difficulties—the personality variable appears largely unnecessary in seeking explanation and understanding of what happened. It would have been remarkable had there not been conflict between the Engineers and the Agricolas. As it was, the cumulative and mutually reinforcing tensions and differences between departments held them further apart and raised the temperature of exchanges higher than might have been the case had fewer factors been operating. Consequently, the differences were clearer than they might otherwise have been.

The manner in which issues between the scheme-level officers of ALDEV and the M.O.W. were worked out reveal that they relied on different alliances. During the early years of the Scheme the ALDEV Management depended heavily on support from the Administration. The D.C., by virtue of the dominant position of the Administration and his chairmanship of the Local Committee, was a powerful figure able in many ways to assist the Management. When, for instance, there was considerable disagreement between the Management and the Engineers over the installation of village water supplies, the D.C. ruled in favour of the Management. The Engineers, for their part, considered that at the Local Committee meetings the Management obtained decisions from the D.C. on matters which should have been their own domestic concern and not a subject for debate at all. Relatively weak as they were on the Scheme because of the close relationship between ALDEV and the most powerful local patron, the Administration, the Engineers relied on their department in Nairobi and the voice of the Hydraulic Engineer on the J.I.C., which they considered the correct forum for deciding all technical matters. Departmental differences thus took on a local-central dimension. In 1957 the Local Committee, dominated as it was by the Administration and the Management, decided to adopt the level basin system advocated by the Agricolas for the red soils. This provoked a powerful protest by the Hydraulic Engineer in the J.I.C. where he questioned not only the decision, which conflicted with the advice of his department, but also the right of the Local Committee to make such a decision at all. The Engineers also sought to establish another form of separation between the Scheme and the J.I.C. Because he had earlier been the peripatetic irrigation adviser, the first Manager sometimes continued to attend J.I.C. meetings. The Resident Engineer, his opposite number on the Scheme, did not, however, attend. The Engineers considered that this gave the Management an unfair advantage, since only one side of an issue which had arisen at scheme level might be presented to the J.I.C. They therefore proposed that either the Manager should not attend J.I.C. meetings or the Resident Engineer should also be present to put his point of view. A balance of alliances and central-local power was not enough in itself; it was also considered necessary to establish and enforce the rules of play.

In these circumstances the authority of the Administration was enhanced. The differences of approach between the Engineers and the Management generated a need for an umpire and arbitrator, and the D.C., in his position as chairman of the Local Committee, was cast for the role. But the Administration was also in a powerful position because of its responsibility for local political matters, which included the activity of land negotiation which was fundamental to the work of the other two departments, since without occupation of land there could be no scheme.

The land and the local people

The local people held a strong claim to the land on which the irrigation system was constructed. The official view from the beginning was that rights to the land should be acquired by government, but the problem on the official side was to know what land was required. The area considered quickly rose from 5,000 acres in June 1953 to 9,700 acres towards the end of that year, and in February 1954 ALDEV applied to the Native Lands Trust Board for the setting aside of 12,000 acres. But nothing happened, perhaps because of uncertainty about the final area that would be required and a sense that other issues were more immediately important.

This rather dilatory approach was possible because of the atmosphere of the Emergency. The Administration was supremely confident of its ability to achieve the changes it sought, and believed in and practiced government through a sort of consent by compulsion. The people of Central Province were like patients under an anaesthetic; the surgical operation of land consolidation, and the grafting on of the Mwea Scheme were carried out while powers of political protest were numbed by the threats and incentives of Emergency administration. It is not surprising therefore, that the minutes of the Embu A.D.C. do not record any dissent to the idea that there should be a scheme. But the Councillors continued their earlier struggles for land rights in the Mwea by openly differing from the Administration in their view of whom the Scheme should be for. They feared more settlement by Kikuyu *ahoi* and wished to displace those who had already established themselves in the Mwea. They held that the *ahoi* had been invited into the District between 1933 and 1947 by the more powerful chiefs of that time against the wish of the local people; that they were subversive; and that they should be returned to their districts of origin. As elsewhere, the Emergency was being used as an opportunity for attempting to settle old land disputes. The Administration's repeated assurances that those resident in the area developed by the Scheme would be given first option to take up plots were weakened in their appeal to the Ndia and Gichugu by the fact that many of those affected were *ahoi*. The Councillors sought to make their agreement to the setting aside of the land conditional on the Scheme being for Ndia, Gichugu and other original inhabitants of Embu District, and opposed the introduction of outsiders. The Administration, however, still saw the Scheme as a means of settling landless Kikuyu from other parts of Central Province, and was only prepared to concede that a small number of local people, namely those whose land had been occupied by the Scheme, should be allowed plots.

One effect of the Scheme was to crystallize, indeed to attract, claims of land rights. In 1956 the Local Committee noted that more and more local people were claiming land rights in the area, and were becoming increasingly reluctant to move off the land they were cultivating as the rice fields advanced. Growing resistance to the expanding boundaries of the *de facto* Scheme was reflected in the theft of survey pegs and other petty irritations. There was probably also some speculative cultivation in areas which were about to be developed with the hope of acquiring settlement or compensation rights.

To solve the land problem, the Administration considered various proposals: a produce cess to be paid to Embu A.D.C.; annual compensation payments to the A.D.C. or to local landowners; payments by other A.D.C.s which would be providing settlers to Embu A.D.C. in recognition of the benefits they were deriving; and the settlement in the Scheme of local right-holders. Such adaptive responses as these might have been agreed had they been introduced early enough. But as the effects of the Emergency wore off the chance slipped by and agreement became more difficult to achieve.

The problems of land negotiation were aggravated from November 1955 onwards by the arrival of Kikuyu potential settlers from Kiambu, Fort Hall and Nyeri Districts. They embodied the Administration's view of the purpose of the Scheme and presented a visible target for local objectors. But other effects of their arrival, though not immediately noticeable, were more far-reaching. They altered the nature of the Scheme and generated new activities and preoccupations which affected departmental alliances. Although the activities associated with physical development continued until mid-1960, the arrival of potential settlers shifted the centre of attention from changing land to managing people.

CHAPTER 5

Ruling the Settlers

The period from the beginning of the settlement of tenants on land in 1957 to the completion of physical development and also of settlement in 1960 is the most complex in the history of the Mwea Scheme. The extent of the changes which took place is suggested by the fact that at the end of 1956 no tenants had been settled and only 589 acres had been prepared for irrigation, whereas at the end of 1960 there were 1,244 tenants on holdings and 5,226 acres of irrigable land had been developed. The period includes the complete range of activities —policy formulation, survey, local negotiation, planning, construction, settlement, production, and organization—which were carried out in creating the Scheme as a largely independent entity. The mesh of activities and events is so interwoven that any separation of strands for the purposes of analysis risks distortion. Nevertheless the arrival of potential settlers in November 1955, and the start of their settlement as tenants in 1957, introduced a new thread into the Scheme around which, as it were, other threads clustered and arranged themselves. These settlers, representing yet another invasion of the Mwea, yet another layer of colonization, altered the attitudes and relationships of the government departments which had preceded them. They stimulated the Administration and the Management into expressing different views of the purposes of the Scheme. They raised the questions of which department could claim to administer them, and of which department should 'own' the Scheme. They provoked the local people into new forms of behaviour towards the Scheme. They also changed the nature of the Scheme by adding a new human and social dimension. It is these processes that are described and discussed in this chapter.

The political and administrative environment

First, however, the political and administrative setting must be taken into account. From 1957 to 1960 political activity was more decisive at the national than at the local level. This was especially the case in Central Province, which differed from the rest of Kenya in being subject to continuing Emergency restrictions. In June 1955 when Africans were again allowed to form local political organizations, those in Central Province were limited to Advisory Councils of nominated loyalists.[1] In March 1957, when elections took place for eight African seats in the Legislative Council, there was only one member for the whole of Central Province and only loyalists were included in the limited franchise. Political activity in Central Province continued to be regarded by the Administration as potentially subversive, and in early 1959 many of the prominent Kikuyu political leaders who were emerging were restricted to their districts.[2] It is symptomatic of the Administration's suspicion of African politics and the extent to which it sought to exercise its authority that as late as 1960 a Chief's order was in force in the Mwea banning songs praising Kenyatta or demanding his release. Indeed, during this period, the Administration had a far greater influence on the development of the Scheme than any African political organization. The posting in of a D.O., and the creation in June 1956 of a new Mwea Division to include the Scheme and also the lower Mwea, reflected both the importance attached by the Administration to the Scheme and the denser staffing of the closer administration introduced to handle the Emergency and its aftermath. Local government, in the A.D.C. at district level and in Locational Councils at location level, was dominated by the Administration and its nominees, and the presence of a Headman and Tribal Policemen in each village made possible the exercise of considerable control over the lives of the people. The Administration was still firmly in command, and the Scheme evolved largely independently of national political changes.

Restrictions were gradually eased and a return to peaceful conditions took place. The bulk of the population was now less preoccupied with survival in civil war than with what were felt to be the necessities of life—food, money, clothing, housing and education. The rehabilitation of detainees, mainly carried out in the Mwea camps, was far more rapid than had been expected. Despite a scare in October 1958 of a revival of Mau Mau[3] there was a generally steady relaxation of tension. As the D.C. Embu put it in his review of 1957:

---

[1] George Bennett, *Kenya, a Political History: the Colonial Period*, Oxford University Press, 1963, p. 138.
[2] Rosberg and Nottingham, op. cit., p. 317.
[3] The movement was called *Kiama kia Muingi* (K.K.M.)—the People's Party. The main significance of K.K.M. in the history of the Scheme was that it enabled the Administration to maintain its powerful position through the over-riding demands of security. Forty-five persons on the Scheme were placed on orders requiring them to report to a Chief each day. K.K.M. is briefly described in Rosberg and Nottingham, op. cit., pp. 306–307.

The District continued to shake off its encumbering Emergency vestments. The unfamiliar, almost forgotten and somewhat shabby garments of day-to-day life required considerable adjustment and, in some instances, the people were not quite sure whether they were wearing their own clothes or had bewilderingly acquired someone else's. A little prodding here and there and much shrugging of shoulders usually in the end resulted in an acceptable fit.[1]

The administrative changes which most affected the Scheme were within the Ministry of Agriculture. In early 1956 a General Manager, Irrigation Development Projects, was appointed. Coming from the Sudan where he had worked on Gezira, having a higher salary than the most senior administrative and agricultural officers, and lacking a clearly defined role among government organizations, it is not surprising that he was to some extent regarded as an interloper. Symbolically his attendance at the J.I.C. was for over a year only by invitation, and only once during his four years as General Manager did he chair the J.I.C., and then only when both the Permanent Secretary of the Ministry of Agriculture and the Director of Agriculture were absent.[2] The General Manager worked from the Ministry of Agriculture, with which the Management of the Scheme became more closely associated. In 1959 the Ministry of Agriculture took over accounting for ALDEV and the Works Supervisors were appointed Assistant Agricultural Officers (A.A.O.s) within the Ministry. When the General Manager left in 1960 and was not replaced, the Scheme Manager worked to the Chief Agriculturalist. The Scheme had become a subunit of the Department of Agriculture and its direct connection with ALDEV ceased.

The introduction of settlers

The selection of settlers, in accordance with the aim of relieving the pressure of landlessness in Kikuyu districts, was carried out by the Administration. The original intention that 'loyalists' should be settled was superseded as it came to be thought that rehabilitated detainees, known as 'whites', would pose a problem when returned to their districts—some because they might not be acceptable to the local loyalists, others because they might lack means of subsistence and backslide into Mau Mau. The first potential settlers to arrive, in November 1955, were landless 'whites' who had been selected by the Administration. Quotas for the three Kikuyu districts of Kiambu, Fort Hall and Nyeri were laid down by the P.C.'s office, and throughout 1956 batches of 'whites' arrived on the Scheme. Some had

---

[1] Embu District Annual Report for 1957, p. 1.
[2] The meeting of 24 March 1959.

volunteered; others were said to have been compelled by their local chiefs. Previous agricultural experience was not considered a relevant qualification. In addition, it was decided early in 1956 to accept local people who could show that they were pre-Emergency right-holders. By the end of 1956 there were 1,418 potential settlers on the Scheme,[1] most of them landless 'whites'.

To the settlers the Scheme did not appear to differ much from an open prison. Some did not have their own clothes and wore the same as the detainees in the camps. Their personal possessions were very limited, in some cases non-existent. They came without their wives and families. They lived in a camp and ate communally. Their work was at first supervised by Prisons *askaris*. There was a muster call twice daily and although they were allowed to move within a seven mile radius they had to be back by 4.00 p.m. each day. It is not surprising that there were complaints and defections, and that the local Ndia, Gichugu and *ahoi* rightholders were reluctant to take up work on the Scheme with a view to becoming tenants. If they did so they feared they would become prisoners, be unable to visit their relatives, and 'serve the whites (Europeans) forever'.[2] The Scheme was perceived as, and indeed in many respects was at this stage, a total institution in which potential settlers were inmates over whom a very high degree of control was exercised by the staff.

On the other hand there were respects in which it was indicated that a different style of life was intended. The settlers were paid Shs.45 a month (later Shs.55) and were employed on road-building, house construction in the villages and field levelling. The possibility that their families could join them was held out as an incentive to encourage hard work on the housing. The morale of the settlers became a preoccupation, particularly of the Administration, which applied pressure both on the Medical Department to allow more rapid occupation of houses, and on the Management to allow settlement on the land.

Much of the difficulty of the induction of settlers resulted from uncertainties about the organization and economy of the Scheme. One basic question which exercised the settlers was security of tenure, but neither the Management nor the Administration could give more than general indications of the tenure system since it had not been worked out. Decision was delayed as to whether settlers should have two, three, four, five or six acres (each figure being tried or proposed at some time) and whether holdings should be on the red or black soils or on some combination of the two. Nor was there an assured cropping policy. In 1956 it was not even known whether rice

[1] Mwea/Tebere Irrigation Scheme (M/T.I.S.) Annual Report for 1957, p. 4.
[2] Phrase used by an ex-detainee in an interview (1966).

would grow on the black soils of Tebere. The devastation wreaked by 'quelea'[1] birds on the 1956 rice crop, though they were subsequently controlled by aerial spraying, cannot have inspired confidence, nor can the fact that one of the first tasks of the tenants in Tebere in 1957 was to uproot the rotting maize and beans crop that had been planted on the black soils of their tenancies.

Settlers were also doubtful whether they could make a good living off the Scheme. Rice was a novel crop to them, and a market for large quantities of it was difficult for them to conceive. Moreover, for them to use it as their staple subsistence food involved a major adjustment. Another source of uncertainty arose over the payment to be made to the Scheme for services provided. Settlers did not know how it would be levied nor how large it would be. The idea of sharecropping canvassed by the first Manager was, after a debate, abandoned in favour of a rent on the grounds that with sharecropping estimation of crops and their collection would pose problems. The rent of £15 per annum per acre proposed by the J.I.C. caused alarm among potential tenants and provoked a battle between the J.I.C. and the Local Committee, which took the view, in a series of proposals, that little or nothing should be paid in the first years, but that rent might rise subsequently. The final compromise—£1 for the first year, £5 for the second and £10 for the third and thereafter—must have encouraged the settlers, but only after a disturbing period of uncertainty.

These problems, coupled with delays in settlement resulting from slow land preparation and uncertainties about crops, led to defections. Despite new arrivals the total number of settlers and tenants dropped from 1,418 at the end of 1956 to 910 at the end of 1957 and 577 at the end of 1958.[2] Of 1957 the General Manager reported that: 'More tenancies could have been given out in the Nguka swamp area, but many settlers decided against taking up a tenancy and wished to be returned to their reserves.'[3] These large-scale departures were ironically a long-term asset to the stability of the Scheme as an institution. Had all the first settlers remained, they could not all have been given 4-acre holdings by the end of 1960, when the Scheme was fully settled with only 1,244 tenants. Not only would those who were left over have presented a problem, but it would have been impossible to hold out the incentive of a chance of settlement to the ex-detainees who remained on as paid labour working for Management and M.O.W. More important, it would have been difficult to absorb the growing local pressure to become tenants. The defections thus freed

[1] *Quelea quelea abyssinica*, the Sudan Dioch, colloquially known as quelea.
[2] M/T.I.S. Annual Reports for 1957 and 1958.
[3] *Department of Agriculture Annual Report 1957*, p. 46.

the Scheme from the burden of the earlier political decision that it was a place where surplus Kikuyu population could be accommodated. Short-term failure permitted long-term adaptability.

The popularity of the Scheme with the settlers rested primarily on its capacity to provide them with a relatively high standard of living. The turning point was the successful rice crop of the 1957 short rains. Early in 1958 it was correctly supposed by the Management that there would be no further difficulty in finding tenants. The following year when they achieved net average returns per four acre tenancy of £109 from the 1958 short rains crop (excluding rice retained for consumption which was considered equivalent to the labour input),[1] tenants appeared to be exceeding the Swynnerton Plan target of £100 plus subsistence for each family.[2] The profitability of the Scheme to the tenant was and remained such a strong incentive that irritations and problems were relatively easily resolved, since the tenants were anxious to go for the main economic chance.

The Administration and the Agricolas: two views of the Scheme

The introduction of settlers and the growing physical presence of the Scheme raised in a pointed way a number of questions about its purpose and organization. These were brought to a head by the proposals of the General Manager in 1956, and again in 1958, for the setting up of an Irrigation Board as a largely autonomous corporation to run irrigation schemes. What sort of organization was the Mwea/Tebere Scheme, and what was it for? Who was to control the settlers? Who was to be able to claim to speak for them? Who was to be the ultimate authority on the Scheme? What department, in fact, was to 'own' it? These issues divided the Administration from the Agricolas. Their differences at central government, provincial, district and scheme levels are so consistent that it is possible without serious distortion to speak of two coherent and contrasting departmental bodies of opinion.

At the most obvious level of interpretation, the Agricolas' interest was in growing crops and in the economic process of which growing crops was a part. To them, the Scheme was to become an estate for them to manage, and the more it became like an estate the stronger and more insistent became their interest in it. Approaching partly from an economic point of view they valued economic self-sufficiency and independence of subsidy, and associated this with satisfaction on

---

[1] M/T.I.S. Quarterly Report for April–June 1959. The short rains crop was harvested in the calendar year following the rains. Tenants retained some rice for family subsistence and also cultivated areas of unirrigated red soil. However, it was probably also necessary for them to buy food.

[2] *A Plan to Intensify the Development of African Agriculture in Kenya*, op. cit., p. 58.

the part of the settlers. As the irrigation adviser wrote in his 1954 report:

> If these schemes are run on sound commercial lines rather than as Government subsidies, the possibilities of a successful partnership between the Government and tenant-farmer might well lead to a greater appreciation and contentment on the part of the tenant cultivator than if he felt he was being subsidised as a political expedience.[1]

He looked forward to a situation in which the Scheme would be able not only to repay capital charges and cover its operating costs but also achieve a good margin of profit. This profit would be employed to develop an internally interdependent economy with pasture under flood irrigation, rice residues being used for animal fodder, and manure from the animals being used on the rice fields. The vision was that of many farmers: a stable, profitable and balanced organization of mixed farming, independent of external assistance.

In order to farm successfully the Agricolas maintained that the Manager should be able to exercise close control over the Scheme and the tenants. In part this was justified on the grounds that the Kikuyu had no previous experience of irrigation and had a recent record of opposition to schemes for agricultural development. During the Emergency it might have been possible to compel the Kikuyu to work efficiently, but trouble was foreseen with the relaxation of Emergency controls. Close control was also considered necessary for technical reasons. The requirements of water control and crop husbandry and the synchronization of agricultural activities made it essential, it was argued, that the Manager should be able to exercise strict disciplinary powers.[2] The Manager himself put it that he should be 'solely and entirely responsible for all happenings on the Scheme'.[3]

The economic independence and close internal controls sought by the Agricolas had implications for the administrative independence of the Scheme and for its relations with its local organizational environment. These issues were brought out by the General Manager in 1956 in his proposals for an Irrigation Board at the centre.[4] This Board would have given him a much greater degree of autonomy, with administrative and financial responsibility for all major irrigation schemes, and would also have established a greater separation of individual schemes from departments of government at the local

---

[1] The Irrigation Adviser's report on his visit to Mwea and Tebere Irrigation Projects (typescript), 21 April 1954, p. 6.
[2] F. A. Brown, op. cit., p. 12.
[3] Letter, Manager to Acting General Manager, 9 April 1958.
[4] Memorandum on the organization of irrigation by the General Manager, April 1956.

level. He saw no need for a local committee of management after the construction stage. Managers of schemes, he felt, should be free to consult representatives of government departments as the need arose, but should not be subject to their control. Government officials at the local level were constantly changing so that they were best left out of management except for consultation where necessary. The relationship of the Mwea Scheme to the Administration was foreseen as similar to that of a large estate in the European settled area, which meant that the Administration would be responsible for little more than law and order and tax collection, since on estates many of the welfare functions of government were normally performed by the Management. In sum, the General Manager wanted irrigation schemes to be organizations separate from their administrative surroundings, and controlled by him from the centre rather than by local committees dominated by the Administration.

The replies which the General Manager's proposals drew from the Administration in the field were characteristically outspoken, and expressed very different views from those of the Agricolas about the purposes of irrigation schemes. These are especially revealing for the two provinces, Rift Valley and Central, where the two principal irrigation schemes, Perkerra[1] and Mwea, were situated. Writing of Perkerra, which had started in somewhat similar circumstances to Mwea, the P.C., Rift Valley Province, said that the General Manager's proposals missed the main point, which was that:

a scheme such as Perkerra, and no doubt others, is *not* a Government Estate for making money, growing produce and employing labour. If it were, then the Central Board would be excellent. But in fact it is a settlement and land rehabilitation scheme ... the Perkerra scheme *must* be considered as only one part of the general rehabilitation of Baringo reserve.[2]

The Special Commissioner, Central Province, wrote of Mwea:

The Mwea-Tebere Scheme is primarily a resettlement project to relieve pressure in the Kikuyu Reserves. Since it is an experiment in irrigation methods on a large scale, we need experts to direct the technical aspects of irrigation policy and management, but ... these schemes are not designed in the long run as self-contained Government estates. The Mwea/Tebere scheme will not work as an

[1] The Perkerra Irrigation Scheme was in a more remote area and in a less developed district (Baringo) than Mwea, and was regarded by local officers of government, especially the Administration, somewhat more parochially than Mwea.
[2] Letter, P.C., Rift Valley Province, to Secretary for African Affairs, 20 June 1956. (P.C.s italics).

irrigated island divorced from the law and order and the other problems of the surrounding district which are the responsibility of the District Commissioner.[1]

In both these cases, the Administration saw the Scheme as part of a wider local environment, an environment which, of course, it largely controlled. Moreover, administrative officers could see the proposals for the Board and for independent management of irrigation schemes as threats to their influence which might lead to the virtual excision of parts of their territories.

The differing views of the Scheme held by the Agricolas and the Administration were reflected in their attitudes to tenant management. However humane their approach might be, the Agricolas necessarily regarded the tenants in one aspect as a component in an agricultural process, as a means to an economic end. The Administration, largely unhampered by economic criteria, set a higher value upon the quality of personal relationships with the tenants. There was good reason for departmental officers referring to D.C.s and D.O.s as 'squires' and 'lords of the manor', since many of them consciously or unconsciously harboured traces of a feudal image of themselves and their job. The D.O. wrote of the young British staff on the Scheme that, though excellent young men, some had a 'lower deck mentality', and that 'if they are to have sole charge of the people in the villages as well as in the field, success will depend upon and demand their ability to adopt a more paternal attitude of landlord than that of foreman of works'.[2]

The statement can be taken straightforwardly as indicating that only suitable staff should hold positions of responsibility. But at another level, it raises the question of what sort of person should handle the tenants, and who should run the Scheme. The Administration's view was that it should be an officer with experience of the Kikuyu as well as knowledge of irrigation and agriculture. It was considered that it might well be that a senior administrative officer, with a full staff of technical officers, would be most suitable. The most effective way to ensure that the Scheme did not become 'an irrigated island' in a reserve might be, in fact, to recolonize it, to reclaim it from the Agricolas and to evolve an organization and a set of relationships more like those of a feudal manor than of a commercial estate.

These different views were bound to give rise to conflict and competition. In 1954 the adviser from Gezira suggested that:

The fact that agricultural interests are paramount in the irrigation

---

[1] Letter, Special Commissioner to Secretary for African Affairs, July 1956.
[2] D.O. Mwea's Report for December 1956, 19 December 1956.

area is likely to cause a certain amount of friction, natives being adepts at playing off one department against another. In anything to do with the efficient running of the Scheme, the agricultural manager should have the last word (within reason), and be given every support by the district and provincial administrations.[1]

But the Administration did not accept that agricultural interests were paramount, and the problems that arose between the Administration and the Agricolas cannot be attributed more than marginally to tenants playing off one department against another, though that did occur. Rather, the tenants were responsible for new problems because their presence generated roles for which organizations and individuals competed. The issues were not just what the Scheme should become, but whose subjects the settlers should be.

Who controls the settlers?

On the Scheme the Administration and the Management competed for the occupation of land, the control of people, and the performance of roles. Since the Administration had some sense of ownership of territory, and of rights to control and represent the people and to perform new governmental roles,[2] it might appear that in so far as the situation was competitive, the Management were aggressors. This would, however, be too simple an interpretation. When the D.O. was posted in the Manager had already been in occupation of the ground for over a year. What is more, through altering and cultivating some of the land, the Management was carrying out activities implying rights of use which were not far from departmental ownership. To the Manager it was the D.O. who appeared as the potential aggressor. At an early stage the Manager argued that the D.O. should not occupy the same territory. He was anxious that the D.O.s area should not coincide with the Scheme area and that his designation should not be 'District Officer to the Mwea-Tebere Scheme'. He would then '. . . more easily retain the point of view of a District Officer and thus run less risk of identifying himself with the Scheme Management, otherwise there is always a possible source of friction, especially when the District Officer is young and enthusiastic'.[3]

These wishes were met. The Mwea Division that was formed in June 1956 included the lower Mwea, as well as the irrigation scheme, and the D.O. who was posted in was a relatively old man with wide

[1] F. A. Brown, op. cit., p. 13.
[2] See pp. 50-52.
[3] Letter, Manager to D.C., 26 June 1956.

experience and whose tact, understanding and lack of aggressive ambition toned down the conflicts between departments.

The rivalry between the Administration and the Management centred at first on the settlers. In harmony with its view of the purposes of the Scheme, the Administration felt that many of the problems which would arise on it would be of a political nature, and that it would be necessary for the D.C., acting largely through his D.O., to play a major part in the administration of the Scheme. The Administration, having recruited the settlers, considered itself responsible for their welfare after arrival and sought to exercise control over the village communities that were set up, and which provided a focus for the interplay of the interests of the Administration and the Management. A village form of settlement was adopted for security reasons, to simplify management, and to avoid the occupation of unnecessarily large areas of irrigable land. Since villages in the rest of Central Province were organized and controlled by the Administration through appointed Headmen, Tribal Police, and village committees, it was natural for the Administration to think in terms of setting up a similar system on the Scheme. But whereas agricultural practices elsewhere were a matter for individual choice, on the Scheme they were to be subject to managerial control. This meant that the total amount of control exercised by government over the people would be greater on the Scheme, and it also meant that the exercise of that control would be divided between the two organizations, the Administration and the Management. How to divide it was not easily resolved.

The first detailed attempt to demarcate responsibilities was made by the D.O. He proposed a dual organization with two distinct lines of supervision to a council of elders in each village.[1] Chairmen could be either from the Administration or from the Management, according to the purpose of the meeting. These councils were to have separate social-political and agricultural functions: directing decent behaviour in general, and furthering the interests of prosperity through good husbandry in particular.[2] Each council was to have two executive officers: a Head Cultivator,[3] paid by the Management, responsible for passing on the Management's orders to the villages and seeing that they were carried out; and a Village Headman,[3] a tenant who would be an Administration nominee, paid a retaining fee, and responsible for receiving 'government orders from the Chief, interpreting them to the people of the village, and seeing that they

[1] D.O. Mwea's proposals for Scheme organization, February 1957.
[2] Ibid.
[3] At this stage the terms Cultivation Supervisor and Village Leader were used, but the titles Head Cultivator and Village Headman were subsequently adopted and are used here to avoid confusion.

were observed. With any Tribal Police who might be posted to the village, the Headman would be responsible for law and order, 'village interior economy', housing, sanitation and educational matters. Both the Head Cultivator and the Village Headman would have power to call their village committee into session on matters for which they were responsible, and would chair that session. The committees were to be established at a public meeting held jointly by the D.O. and an ALDEV officer.

The Manager, however, sought a higher degree of control than he considered these proposals to allow. He feared that the degree of democracy envisaged would lead to delays in cultivation matters. The system might be appropriate for control of cultivation for a reserve, but not for 'an estate where Block Managers and Inspectors would give instructions direct to their Head Cultivators'.[1] Where, for example, spraying had to be carried out against a blight, the Management must have power to employ paid labour and debit the cost to the tenant concerned, without the delays of deliberation in a village council. The Management, in fact, sought to establish a comprehensive administration under its own control. In notes for a meeting with prospective tenants, it was stated that: 'Tenants should realize that the Scheme Management is their Landlord, and look to their Block Inspector for guidance in all matters both personal and cultivational through the medium of the Village Committee.'[2]

Applications for passes to visit other areas 'and personal problems of every kind' were to be brought in the same way to Management staff. The D.O. might attend meetings of villages or village committees 'at the request of the Management'. The benefits of irrigation were to be explained to the tenants, and an attempt was made to encourage the tenants to look towards the Management and not to the Administration for guidance and help.[3]

This degree of autonomy for the Management was, however, impossible in the conditions of the time. The D.O. and the Chief had a legal as well as a customary power to call village or other meetings, and would not have accepted that their attendance was subject to a request from the Management. Moreover, in practice many of the matters raised to the Management by the tenants had to be referred to the D.O. for decision or action. Only the D.O. could deal with queries about poll tax, or questions such as whether settlers could bring their parents or their second wives to the Scheme, whether Headmen could be paid more, whether more primary teachers could

---

[1] The Manager's comments on the D.O.s proposals, 12 February 1957.
[2] Undated notes probably written in May 1957, in Scheme file Tenants Liaison Council 1957–1959, at folio 1.
[3] Ibid.

be provided for schools, or whether Kikuyu settlers could receive trading licences. Further, the Management was constrained to accept co-ordination of labour demands made on the tenants since otherwise an Administration Headman might call out a village on communal labour on a day when they were required to cultivate. In these circumstances the settlers must have found it difficult to know who was in charge of the Scheme, and it would have been unnatural had they not sometimes, when rebuffed by the Management, taken their case to the Administration, especially since they brought with them to the Scheme a conditioning, reinforced by the Emergency, of looking to the administrative officer as the ultimate authority in all government matters. In addition, although the Management might establish their supremacy in certain spheres, the control of the outside environment by the Administration, most notably through the issue of movement passes, remained undisputed and continually leaked, as it were, into the far from water-tight organization of the Management on the Scheme.

The Management was, however, able to establish a new formal means of communication with the settlers. In its May 1957 meeting the Local Committee agreed to the setting up of a Tenants' Liaison Council to include two representatives from each village and to meet once a month with the Management. No mention was made of attendance by a representative of the Administration. That the minutes of meetings of this Council do not appear to have been circulated outside the staff of the Management may be an indication of a desire to make the meetings an internal Management affair. It is not surprising that in January 1958 the D.O. asked if he could have notice of meetings so that he could attend where practicable. However, attendance by representatives of the Administration was rarely recorded. Through this Tenants' Liaison Council, a body which grew in size as the number of villages increased, the Management thus established a largely exclusive channel of communication with the tenants. To the tenants it provided a forum for voicing complaints and making requests and suggestions. To the Management it provided a means of propagating information and, more significantly, a sounding board, a constituency, an institution which could give some support to a claim to be able to speak on behalf of the tenants and to be in authority on their attitudes and problems.

Indeed, who could claim to 'know' the tenants was an important, if not explicit, issue related to the legitimacy of the Administration vis-à-vis the Management. The Administration traditionally regarded itself as the judge of political matters, including assessments of the people's probable reactions to new measures. In 1957, however, the Management successfully challenged their authority over one such question. When land preparation was going slowly, the Management proposed employing settlers on it and paying them by piecework. The Administration held that the settlers would not agree to this arrangement. After considerable discussion, the change was made and proved a success, with an improvement in settler morale and an increase in the rate of land preparation. The Management had effectively set itself up as a rival authority to the Administration on probable settler responses.

The control of settlers was made difficult for the Management by the legal position. In the autocratic atmosphere of the Emergency, government officers could take action which they would have described as 'disciplinary' without undue concern for legal niceties. But by 1958 unofficial fines and confiscations and the 'goat-bags' into which they were placed were becoming less acceptable. In November 1958 it was suggested that Managers of irrigation schemes should be gazetted as magistrates who could apply an Irrigation Ordinance or an A.D.C. by-law, a suggestion unlikely to find favour with the Administration which elsewhere controlled and performed judicial functions at the lower levels. A chief's order was in force covering damage to irrigation works, but it did not provide a means of prosecution for minor delinquencies such as water offences. Even when Irrigation Rules had been agreed in 1959 they could not be enforced until December 1960, by which time the land of the Scheme had finally been set aside and the tenants had signed the Rules.[1] In the meantime, a mixture of threats, persuasion, and unofficial sanctions based upon the withholding of factors in the economic process controlled by the Management—for instance, water for irrigation, or crop advances—had to be used. The ultimate sanction of eviction, partly as a result of its political implications, was at first *de facto*, and later *de jure*,[2] vested in the Local Committee, and therefore not freely at the disposal of the Manager. Until the tenants had signed the Rules the Management's control of them was based upon weak foundations and depended partly upon the goodwill and support of the Administration, since the exercise of an unofficial sanction might always create a political problem which only the Administration could solve.

Who manages the Scheme?

In 1957 uncertainty, indeed confusion, about responsibilities and roles covered not only the ill-defined boundary between the Adminis-

---

[1] Mwea Irrigation Settlement (M.I.S.) Annual Report, 1960, p. 5.

[2] *The Trust Land (Irrigation Areas) Rules, 1962*, Legal Notice No. 535 of 1962, section 22 (3).

tration and the Management over control of settlers, but also the unstable and at times bad-tempered division of construction activities between the Management and the M.O.W. In May 1957 the D.C. wrote a memorandum about 'the large and looming question of control of the Scheme'.[1] He pointed out that the three departments involved, the Administration, the Ministry of Agriculture and the M.O.W., often overlapped in their work. Africans working on the Scheme naturally looked to the Administration for a solution to their personal problems, and frequently brought complaints to the D.O. when they should be directed to the Management. He continued:

> Either through ignorance or design, orders given by one Department can be and are unfortunately inadvertently countermanded, sometimes duplicated, by another and the resulting chaos in the employee's mind is considerable. This of course leads to inefficiency in the labour force and possible inter-departmental friction.
>
> A high level directive is urgently required, which must set out a categoric division of responsibility on the running of the Scheme; or an overall commander must be established to whom all other departmental officers working on the Scheme must be subordinated.[2]

There can be little doubt that any overall commander he may have had in mind would have been an administrative officer. Indeed, it was easy to transfer to the Scheme the model of the administrative officer as arbitrator, as the chairman who holds the ring between technical officers with opposed points of view. The issue that was emerging went beyond control of settlers to control and co-ordination of government departments and management of the Scheme itself.

In the latter half of 1957 and most of 1958 there was a prolonged and complicated debate at all levels of government from the Scheme upwards about a suitable form of organization and control for irrigation schemes in general, and Mwea in particular. Discussion took place within departments, between departments and between committees. The central issue was between the Administration and the Agricolas over who should manage the Scheme and how roles on the Scheme should be divided between them.

It would be tedious and unedifying to describe in detail the course of the debate,[3] though the outlines are of interest. The Administra-

[1] Memorandum by D.C., Embu, dated May 1957.
[2] Ibid.
[3] It *was* tedious and unedifying, as the reader will discover if he cares to consult R. J. H. Chambers, 'The Organization of Settlement Schemes: a Comparative Study of some Settlement Schemes in Anglophone Africa, with Special Reference to the Mwea Irrigation Settlement, Kenya', Ph.D. thesis, Manchester University, 1967, pp. 200–206.

tion sought to consolidate its control largely through the Local Committee. As the respective functions of the Local Committee and the J.I.C. had never been officially laid down, there was scope for interpretation. Both committees had started with three members—one agriculturalist, one administrative officer and one engineer. The J.I.C., with the Director of Agriculture in the chair, had tended to be technical in its concerns, while the Local Committee, with the D.C. in the chair, had been concerned somewhat more with administrative matters. Both Committees expanded rapidly, and as they did so the three main departments increased their representation, in general more or less equally.[1] In practice the Local Committee took a strong independent line, often recording 'decisions', reflecting both the confident authority of the D.C. and the fact that until 1958 it was mainly through the Local Committee that the Administration sought to direct the Scheme.

Matters were brought to a head in March 1958 by proposals issued by the Secretary for Agriculture, himself an administrative officer. These suggested giving the D.O. authority to select settlers, to

[1] Both Committees were frequently attended by non-members 'by invitation'. The expansion of official membership was as follows:

Joint Irrigation Committee

Date added	Official
June 1954	Director of Agriculture (Chairman)
June 1954	Executive Officer, ALDEV (an administrative officer)
June 1954	Hydraulic Engineer, M.O.W.
June 1954	The Government Soil Chemist
(various)	(intermittent irrigation advisers)
February 1956	The General Manager, Irrigation Development Projects
July 1957	Representative of the Ministry of African Affairs (an administrative officer)

The J.I.C. met in Nairobi regularly almost every month from June 1956 until the end of 1961.

Local Committee

Date added	Official
August 1955	District Commissioner (Chairman)
August 1955	The Manager
August 1955	The Resident Engineer
March 1956	Agricultural Officer in charge of the Research Station
March 1956	The General Manager, Irrigation Development Projects
March 1956	The Medical Officer of Health, Embu District
May 1956	The Assistant Manager
May 1956	The District Officer, Mwea Division
May 1957	An African Representative (an Ndia Chief)
May 1957	The Provincial Agricultural Officer

The Local Committee, after meeting sporadically in 1955 and 1956, settled down to monthly meetings from 1957 to the end of 1960.

organize and run all tenants' committees, to co-ordinate all aspects of the Scheme and to administer the Africans who lived on it. The Manager would be responsible for the reception and housing of settlers, labour allocations, and agricultural and economic matters. General policy would be decided by the Local Committee which would refer technical problems to the J.I.C. The implication was that the D.O. would virtually run the Scheme. Indeed, at this time a separate proposal was being canvassed that a D.O. should be put in as Manager.

The Agricolas argued back hard, setting out their views of the Scheme and claiming their right to run it. In the words of one:

> I do not believe that the Mwea-Tebere Irrigation Scheme should be regarded as a Division in the Reserve in that it is a specialized and separate entity into which Government has sunk, and continues to sink, large amounts of capital. The dealings are with tenants and labour, eventually becoming all tenants which demand a different administrative and sociological approach to peasant farmers as in a Reserve. Whosoever co-ordinates all aspects of the Scheme and administers the Africans who live on it must control and direct all aspects of the Scheme. ... Divided responsibilities have so far resulted in order, counter-order, disorder, with consequent wastage of time and money. Without having gone into details, I would estimate that *over £50,000 has been wasted* as a result of lack of a coherent overall plan. It is only when one man is in charge of the Scheme that a plan can be produced.[1]

The Agricolas regarded as an insult to the Manager the implication that a D.O. on a part-time basis could do his job for him. The Manager would have responsibility without authority. Provoked, they raised their demands. The Manager, it was argued, should select his own settlers, and the specialized services of the M.O.W. and the Administration should be decentralized so that their officers on the Scheme would become responsible to the Manager.

At the meeting of the Local Committee which considered the proposals the Administration was perhaps weakened by the fact that the chairman was an acting D.C., and by the presence of three Agricolas, including the Provincial Agricultural Officer. But whatever the explanation, the Committee agreed on a series of compromises which went far towards meeting the wishes of the Agricolas. The final selection of settlers and labourers was to be the responsibility of the Manager from those pre-selected by the Administration. Some tenants' committees were to be organized by the Manager, though

---

[1] Typescript observations in reply to the Secretary for Agriculture's circular of 27 March 1958, in Scheme file Scheme Organization.

the D.O. was to have free access to them. It was implied that managerial and administrative matters could be kept apart in different tenant committees even if the different committees involved the same people. The Agricolas thus repulsed the Administration's moves towards taking over greater control of the Scheme.

Although these agreements in the Local Committee proved decisive, a paper war crackled on over the membership and authority of the Local Committee. The Agricolas favoured a purely advisory committee. The Administration argued strongly that the committee should direct the Scheme. The Administration carried the day with its proposal that evictions from the Scheme should be subject to the committee's approval, but at a ministerial level it was finally agreed in 1959 that local irrigation committees should be only advisory. The formal legitimacy of any claim by the Administration to direct the Scheme was thus removed.

The new position was demonstrated in late 1959 and early 1960 over a question of Scheme staffing. The D.C., as chairman of the Local Committee, pushed through cuts in staff establishment, including the abolition of the post of Transport and Maintenance Officer and the deletion of an electrician; but these cuts were later successfully resisted by the Manager who could appeal upwards in the Department of Agriculture, which was now more closely associated with the Scheme. He could also, in the last analysis, decline to take what was, after all, only advice. The *de jure* autonomy of the Management, in ways like these, became gradually more and more recognized in practice. The Management gradually consolidated control of what went on within the Scheme and progressively excluded the Administration from matters concerning Scheme organization. The Administration was, however, left with the functions of negotiating and mediating on behalf of the Scheme with its political and organizational environment.

The land and the local people

Physical expansion in the form of irrigation works and fields, and human expansion through the introduction of Kikuyu settlers, exacerbated the difficulties between the Scheme and its local human environment. In 1959 and 1960 the attention of the Administration was diverted from questions of control of the settlers and of the Scheme by the explosive issues of the occupation of land by the Scheme and the settlement of Kikuyu on it. The D.C. saw himself as a sort of honest broker between the Management and the local people. They, the Ndia and Gichugu, had never accepted the original concept of the Scheme as a place for settlement of landless Kikuyu and now objected bitterly to those who had been settled and demanded their eviction. This hostility was heightened when, from 1958 onwards, it was realized that land consolidation in Embu District would create a local landless class desiring settlement. The elders resolutely refused to agree to the setting aside of the land since they believed that once this had taken place, government would have a free hand to carry out its original intentions. The issue raised very strong feelings. The Administration considered it a canker in the body politic of the District, and reported that the presence of the *ahoi* on the Scheme was taken to be a dagger directed at the heart of the Embu peoples, resulting in 'bitter conflict with Government'.[1]

Opposition took numerous forms: passionate speeches at meetings, boycotts of meetings, refusals by elders to walk proposed boundaries, a walk-out by the elected members of the A.D.C. in protest against the Chairman's refusal to allow a Council visit to the Scheme, a letter from the elected members to the Governor, and finally deputations to the Native Lands Trust Board and to the Governor. At a local level physical opposition was encountered in the continued removal of survey marks and in opposition to the M.O.W.'s work of development; as a result of this the D.O. had to intervene. On top of these diplomatic difficulties, interest in land ownership was sharpened by the process of land consolidation as this spread down from the higher potential areas into the Mwea, and both consolidation and the negotiation of the Scheme's boundary were delayed by the need for agreement on a frontier between the two systems of tenure. None of this, however, prevented a very determined, indeed almost desperate, Administration from forcing through gazettement of the Scheme's land in 1960.[2]

To meet this political pressure, and to achieve stability and security for the Scheme, a number of adaptive responses were made. The suggested area was severely reduced from 40,000 acres to 15,000, excluding some land that was irrigable, and allowing a deep salient of non-scheme on the red soils of Tebere, separating the two main blocks of Tebere and Nguka.[3] The Administration also agreed that land consolidation, which provided individual title, should extend further into the Mwea than was desirable on purely agricultural grounds, in an attempt to draw off some of the pressure on the land suitable for the Scheme. The idea of paying rent to the owners, put forward in 1958 when the Scheme's continuation did not seem completely assured, was abandoned for outright compensation by *ex gratia* payment at a rate of Shs.20 an acre. Acceptance of the payment was, however, repeatedly refused by the clan elders, who no doubt

---

[1] Embu District Annual Reports for 1959 and 1960.
[2] In Gazette Notice 3099 of 5 July 1960.
[3] See map on p. 133.

remembered Chief Njega's oath, and the money was eventually returned to the Treasury in 1963. Another response, supported by the D.C., was to impose an A.D.C. cess on rice, comparable to cesses on other crops grown in the District, but this was opposed by the Agricolas; its eventual acceptance was negotiated on the basis that the A.D.C. would finance the medical services on the Scheme. By far the most important adaptation was, however, the progressive acceptance that recruitment of settlers would be limited to local people, which in the long run meant Ndia and Gichugu. At first this took the form of an acceptance in principle that those already living on the land (which included some *ahoi*) and those with land rights in the area taken over, would be eligible for settlement, but this shifted to agreement that so long as there were suitably qualified Ndia or Gichugu who wanted to be settled, outsiders would not be allowed in. The force of these concessions was at first weakened by the reluctance of local people to settle on the Scheme. In 1959 it was decided to select 200, but only 41 came forward. But the force and effect of local protests is shown by the fact that some 500 Kikuyu labourers who worked on the Scheme during construction, and who had been given some reason to hope that they might receive plots, were removed when their work finished in mid-1960. At the end of 1960, of the 1,244 tenants, 568 were from Embu District, and 676 were *ahoi* from Kiambu, Fort Hall and Nyeri Districts,[1] and it had been agreed that future recruitment would be limited to Embu district.

By the end of July 1960 the major frontier problems of creating the Scheme had been resolved. The land had been set aside and much of the boundary had been marked with conspicuous cairns. The physical expansion of the Scheme had been halted. The M.O.W. had completed the handover to the Management of irrigation works and fields. The Administration had in effect handed over the settlers to the Management to be tenants on the Scheme. With the land gazetted, the way was open for putting tenant discipline on a secure legal footing. Adaptations had been made to local political pressures. The Scheme had become a geographical area, a legal entity, a community of tenants, and an organization, all with accepted boundaries to which no immediate change was intended. Freed of major disputes the Management was able to turn inwards and concentrate on what went on within the Scheme, and in particular on the organizing of production.

[1] M.I.S. Annual Report 1960, p. 4.

CHAPTER 6

Organizing Production and Growth, 1960–1966

The completion of physical development by 30 June 1960 was recognized as a watershed in the history of the Scheme, dividing a period of construction from one of consolidation and organization.[1] Certainly it had some dramatic consequences. The M.O.W. team, with its senior staff of seven, its large labour force, and its heavy machinery, left the Scheme. The Management's own staff was reduced by the departure of the Labour Officer, the Assistant Accountant, and two out of the previous five A.A.O.s.[2] The Management, freed from development problems arising out of M.O.W. activity and progressively released from the close attentions of the Administration, was able to concentrate its energies on the organization of the Scheme, the streamlining of its processes and the integration of its constituent parts. A major change of emphasis was recognized to be taking place. Efficiency within the Scheme was now all-important. In accord with this trend, the Manager suggested to the Local Committee that the title of the Scheme should be altered. Because of local antipathy to the word 'scheme', and also for the sake of brevity, it was agreed that the name should become the Mwea Irrigation Settlement.

Effects of external political and administrative changes

In order to understand the activities and events involved in organizing production and growth in the period 1960 to 1966, it is necessary first to take into account those political and administrative changes outside the Settlement which impinged upon its development. At the national level there was a political transformation. The first Lancaster House Conference of January 1960, at which it was made clear that independence for Kenya under an African majority was an early objective of the British Government, marked a turning

---

[1] *Department of Agriculture Annual Report 1960*, p. 50.
[2] M.I.S. Annual Report 1960, p. 1.

point. In the economic field there was a loss of business confidence in Kenya and an outflow of capital. In the political field there was an intensification of activity, with the formation of KANU[1] and KADU[2] as the two principal political parties, and mounting fears of domination of one tribe by another which were partially allayed by agreement on a complex, decentralized regional constitution. Independence under a KANU government came in December 1963 and was followed by the dismantling of the regional constitution and a tendency towards a one party government.

These major changes had local political effects, which were the more noticeable by virtue of the contrast between the earlier 'discipline' of the Emergency and the relatively free political activity and discussion which now took place. In Embu District the D.C. considered that 1961 would probably be remembered in future years as a year in which politics first really gripped the District. Political activity was at first in opposition to colonial government, from the Governor down to the Headman and Tribal Policeman, and expressed itself in boycotts of elections to local government in Central Province in 1961, and in excitement at the expectation of rich rewards from independence. But, simultaneously, tribal consciousness became more acute and past land boundaries were called in question. The Mwea Irrigation Settlement was affected through a greater sense of insecurity among Kikuyu tenants following proposals to redraw boundaries. Up to this time the Mwea had lain in the mainly Kikuyu Central Province, and questions of its administration had concerned which district within the Province should include it. The issue now was more serious, concerning which of the new regions the Mwea should be placed in, whether with the Kikuyu in the new Central Region or with the Embu, Mbere and Kamba in the new Eastern Region. Some of the Kikuyu tenants feared dispossession of their tenancies if the Settlement went to the Eastern Region, and appealed for title to their land or resettlement elsewhere. A Regional Boundaries Commission, however, partitioned Embu District so that the Gichugu and Ndia formed a new Kirinyaga District, which included the Settlement, and became part of the Central Region while the Embu and Mbere joined the new Eastern Region,[3] retaining the lower Mwea area outside the land required for the Settlement. This redrawing of boundaries, while tacitly acknowledging the claim of the Gichugu and Ndia to the Settlement and the upper Mwea, recognized the lower Mwea as a Kamba and

[1] The Kenya African National Union.
[2] The Kenya African Democratic Union.
[3] *Kenya: Report of the Regional Boundaries Commission, detailed description of boundaries*, Cmnd. 1899–1, H.M.S.O., London, September 1963, pp. 9–10.

Mbere sphere or influence and occupation. In what appeared to be a final partitioning of the old Mwea, it was now divided not just between tribal spheres of influence, but between regions.

At the same time tribal pressures were exerted on the Settlement to recruit tenants and junior staff only from local, that is, Ndia and Gichugu, sources. Before independence, pressures for Africanization and for more representative institutions also affected the Settlement and were reflected in staff and tenant organizations. However, a desire on the part of traditional elders to take over the Settlement that was current in 1960 and 1961 was never a credible or real threat; and when in December 1960 there was a scare that tenants would refuse to sign new tenancy agreements, it proved unfounded. Moreover, there is no indication that tenants worked less hard or less well as a result of anti-colonial sentiments or of a belief that national independence would mean an easier life with fewer restrictions and greater rewards. Both before and after independence the Settlement appeared partially insulated from political activities and pressures.

For the Settlement the main influence of political changes was indirect, through their impact on the field civil service. This was profound and had its major and earliest effect upon the Administration.[1] The emergence of political parties with strong popular support directly challenged the claim of the Administration to a primacy based upon its constituency with the local people and severely weakened the legitimacy of its claim to be able to speak for them. Political meetings were attended by thousands; meetings called by the Administration by tens, or hundreds, where they were not boycotted. Further the Administration, both because its political role made it an obvious target and because the non-technical nature of its duties made replacement relatively easy, was the first department in which there was widespread replacement of expatriates by Africans. The most immediate result was a sharp decrease in continuity of staffing.[2] As African officers were sent on training courses to the Kenya Institute of Administration or to Britain, and as expatriates were progressively retired to make way for them, chain reactions of postings were set off. Towards the end of 1963, D.C.s had held their rank on average for only seven months, and had served in their current districts for only four months. Moreover, the new African D.C.s and D.O.s usually had the disadvantage of

---

[1] These changes are described in greater detail in Cherry Gertzel, 'The Provincial Administration in Kenya', *Journal of Commonwealth Political Studies*, Vol. 4, No. 4, November 1966.

[2] The loss of continuity associated with Africanization is a neglected aspect of change in African civil services. For discontinuities relating to the Settlement, see Table 6.1 on p. 110.

shorter experience of administration than their European predecessors. On top of these difficulties a new decentralized system of regional government was partially introduced, and there was widespread uncertainty among African administrative officers about career prospects, powers, roles and responsibilities, with a strong tendency to look to Nairobi as source of security and promotion in a fluid and unstable situation such as their predecessors had never had to face.

In these conditions, there was a shrinking in the scope of the activities of the Administration. It was all that administrative officers could do to keep their heads above water with the commitments they could not avoid: among others the organization of two national elections and a population census, the implementation of regionalism and the adjustment of regional and district boundaries. In these circumstances, departmental officers were able to break away from that degree of control that the Administration had previously exercised. At the field level departmental jealousies directed towards the Administration ceased to be a serious issue. Departmental officers, who were still usually expatriates, welcomed a new role assumed by administrative officers, who were now usually Africans. Earlier, the administrative officer had sought to represent the civil service to the people and the people to the civil service. He now became more of a broker and interpreter between the departmental officer and the politician. It was now to the politician that he represented the civil service, while to the civil service he represented and explained the politician.

The Administration's relationship with the Settlement was affected by these changes. D.C.s subject to rapid posting did not have time, or perhaps inclination, to do more than deal with pressing problems as they arose. Moreover, reductions of staff weakened the Administration's presence. In 1961, as part of a post-Emergency rundown on Administration staff, the 18 Headmen on the Settlement were reduced to three. As experienced administrative officers departed, those remaining were concentrated strategically at headquarters, and the staff in divisions were fewer and less experienced than before. However, after the dismantling of the regional constitution in 1964 the Administration began to regain responsibilities and move back towards its previous primacy. It was transferred to the President's Office and P.C.s and D.C.s became recognized as the President's personal representatives in their areas, taking the salute at parades on national days, and proclaiming official statements of Government policy. But by mid-1966 there was little indication that this had resulted in any change of attitude towards the Settlement which had come to be regarded as a separate entity within the new

106

Kirinyaga District, coming under the Manager for most administrative purposes. During the disruptions of the Administration from 1960 to 1965 the Management had been able to achieve, and then maintain, a high degree of independence from local civil service controls. Far from the Manager fearing domination by the Administration it had become usual for him to go to the D.C. when he needed his assistance.

Political and administrative changes also had a generally unsettling influence on the staffs of irrigation schemes. On Mwea, as long as the senior staff were all European (as they were until November 1962 when an African A.A.O. was appointed[1]) and the junior staff all African, there was a coincidence of the attitudes and problems of both racial community and staff position. This does not appear to have impeded the functioning of the Scheme, but it did determine a differing impact of external factors on senior and junior staff.

The senior European staff, like other European civil servants in Kenya, felt threatened by the political changes which were expected after the First Lancaster House Conference. It appeared unlikely that they could expect a career in irrigation in Kenya. To the third Manager, who was on Overseas Civil Service terms of service and therefore eligible for full compensation for loss of career, this was a less serious threat than to the other senior staff who, as non-designated officers, were not so eligible. To some, also, the prospect of working under an independent African government may not have appeared attractive. Whatever the reasons, however, 1962 was a year of remarkable staff movement, with discontinuity in all six senior staff posts.[2] Two of the A.A.O.s resigned while on leave and further disruptions were caused by other staff taking long leave. Most serious of all, the Workshop Manager resigned, and apart from temporary expedients could not be replaced until 1966. Before these resignations and losses the average length of time an officer had been on his job was 40 months; after them it was only 20 months. This was, however, before the accelerated postings of Africanization had begun. The rate of turnover in the first African A.A.O.s was striking. The first four who were posted to the Settlement remained there for only short periods (the longest being 14 months) before being transferred elsewhere, while training courses abroad caused further disruption. Nor, after the retirement of the third Manager in 1964, was there effective continuity in the post of Manager. The fourth Manager, who had previously been an A.A.O. on the Settle-

---

[1] M.I.S. Annual Report 1962, p. 1.
[2] The principal sources for this paragraph are the M.I.S. Annual Reports for 1960 to 1965.

ment, took over in 1964 but retired on the expiry of his contract the following year. Indeed, with Africanization and de-Europeanization operating concurrently the rate of turnover in 1965 was more hectic than ever: at the end of the year it was reported that half of the senior staff had been in their jobs for less than six months. At this stage a further factor began to operate as recruiting tapped an international catchment. In 1965 under technical assistance an Italian Manager took over, to be replaced a year later by a Dutchman recruited direct by the Kenya Government. In addition British, German and American Peace Corps volunteers spent periods on the Settlement performing a variety of useful tasks, including survey work, accounting and assisting village construction, but their contributions were limited by short contracts. Although in 1966 and 1967 there were signs of improvement, the discontinuities which earlier had resulted from political change were to some extent being perpetuated by the short-term nature of technical assistance.

Junior staff were similarly unsettled, though the rate of turnover was much less. There was local pressure to dispense with alien Kikuyu staff and to replace them with Gichugu and Ndia, but although this influenced recruiting practice there is no indication that it led to the discharge of any non-local staff. More serious were the real and imagined opportunities for further education and better jobs elsewhere, which were associated with Africanization and the recruiting campaigns of commercial and para-statal organizations. Like African senior officers, the junior staff were anxious to make the most of their chances where they could and while they lasted. Where opportunity was seized it might mean, as it did in 1963 for one of the two key junior officers on the Settlement, obtaining a job in the private sector at twice the salary, with assured promotion and training prospects and generous fringe benefits.[1] Where such opportunities were not seized, it could mean a sense of frustration and failure adversely affecting morale.

In 1966, staffing matters were complicated by the introduction of the National Irrigation Board, under which terms of service were to be revised. Although the long-term effect was intended and expected to be greater stability, the short-term effect was to continue the uncertain and unsettled conditions which had limited staff continuity and sapped morale during the previous six years.

Between 1960 and 1967 major changes took place in the organization of irrigation in central government. At first control over irrigation schemes was weakened. Although the Ministry of Agriculture was under less pressure than the Administration to implement rapid Africanization, the pace and diversity of activity it was called upon

[1] M.I.S. Annual Report 1963, p. 2.

to carry out drew its attention away from irrigation. There was some uncertainty about the organization and control of irrigation, and the supervision of schemes was less close. This was partly because the General Manager did not renew his contract and was not replaced; and the Chief Agriculturalist, who took over the General Manager's responsibilities, had much else besides irrigation to occupy him. Partly, too, this withdrawal was a response to the stage of development of the three main irrigation schemes (Mwea, Perkerra and Tana), Mwea at least justifying some decentralization to the Management. In central government political pressures and Africanization do not appear to have had much effect at first. But the J.I.C.'s meetings became less frequent, and from 1962 onwards its decline can be attributed to disorganization originating in administrative and political changes; to the disruption and discontinuity involved when irrigation was placed for a time in the new Ministry of Settlement and Water Resources in April 1962; and to rapid postings as expatriates moved out of and Africans moved into the top posts in the Ministry of Agriculture. The J.I.C.'s recorded meetings declined from 10 in 1960 to 6 in 1961, 4 in 1962, 2 in 1963 and a final 3 in 1964. Fresh proposals for an irrigation board were abandoned in view of the press of legislation and the belief, in May 1963, that under regional government each irrigation scheme would come under the regional authority rather than the central government. In these circumstances there could be no stable central organization to oversee irrigation.

The position of the Local Committee was at first not clear. In October 1960 the view was taken in the Ministry of Agriculture that now that Mwea was established economically it should pass over to routine administration, and that the Local Committee should lapse and be replaced by the District Agricultural Committee, with the Manager working through the P.A.O. to the Chief Agriculturalist in technical and financial matters. In spite of the resulting uncertainty about its future the Local Committee continued to meet, but every two months instead of every month. The D.C. remained the undisputed chairman, but both the role and the composition of the Committee changed in most interesting ways. At first it continued to decide issues. For example, in 1960 it 'decided' that there should be no more than three shops in each village. But its merely advisory nature had been made explicit in the Rules of 1959[1] and in the revised Rules of 1962 (which made it 'responsible for advising the Manager on the general administration of the area in accordance with Government policy'[2]) and was increasingly reflected in its

[1] *The Native Lands (Irrigation Areas) Rules, 1959*, Legal Notice 410/59.
[2] *The Trust Lands (Irrigation Areas) Rules, 1962*, Legal Notice 535/62, para. 3 (1).

109

minutes, which in 1964 and 1965 tended to 'note' matters rather than to decide. In this respect it expressed the changing balance of authority on internal Settlement matters between the D.C.s, who arrived and departed with bewildering speed, particularly in 1963, and the third Manager who remained from late 1959 until 1964.

TABLE 6.1 *The Mwea Scheme: Continuity of District Commissioners and Scheme Managers, 1956—1966*

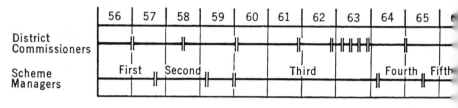

‖ = Change in holder of post

Note: In 1959 the Assistant Manager acted as Manager between the second and thi Managers.

The function of the Committee also shifted, however, from the resolution of conflicts between departments to providing a forum for local political interests. In 1961, at the suggestion of the Manager, the J.I.C. agreed that two tenant representatives might become members of the Local Committee. In 1962 the D.C. secured membership for a representative of the A.D.C. and a member of the A.D.C's professional staff. In 1963, under regional government, the Regional Assembly Member for the area was also co-opted. For its members the Committee became a source of information about policy and for the Manager it became a sounding board for ideas. Its one major executive power was to confirm and approve evictions recommended by the Manager, thus protecting him from the ultimate responsibility for his most sensitive acts.[1] But its main function, from his point of view, was to provide an exchange in which political issues could be discussed and resolved by the D.C. who was now a shield for the Settlement—a buffer between it and its political environment—enabling the Manager to pursue what he felt to be his technical and non-political work undisturbed.

After independence in December 1963, with the gradual decrease in the rate of turnover of civil servants, the issue of the Irrigation Board was again raised. One reason was the steady advance of

---

[1] *Legal Notice 535/62*, paras 8(3) and 22(3).

surveys and proposals for large-scale irrigation on the Tana River, which would have required a special organization in line with practice elsewhere in the world. Another was the less suffocating atmosphere of a government freed from the effective veto of the Administration. In the years immediately after independence this was replaced by a more permissive political leadership prepared, indeed anxious, to innovate, and willing to listen to any civil service entrepreneur with a good proposal he was prepared energetically to promote. An essential factor was the presence of the third Manager who, having left Mwea in 1964, was recruited under the auspices of the United Nations and returned as Officer in charge of National Irrigation. With full support from the Ministry of Agriculture he was largely responsible for the setting up, in 1966, of a National Irrigation Board,[1] of which he became the first General Manager. The Board was a corporation under the oversight of the Minister for Agriculture, but with a high degree of autonomy in matters of accounting and management. The Irrigation Act of 1966 which set it up included a requirement that an advisory committee should be appointed for each national irrigation scheme,[2] which permitted the Local Committee to continue much as before. Indeed, the Board seemed more likely to affect the internal organization of the Settlement than its relationships with its local political and administrative environment.

Organizing the productive process[3]

After the withdrawal of the M.O.W. presence at the end of construction, and with the Administration engaged on Africanization and survival in unpredictable conditions, the Management was freer to concentrate on its main professional interest, the growing of crops. It would be misleading to suggest that the psychological satisfactions of administrative autonomy, particularly with an organization and process as susceptible to detailed control as the Mwea Irrigation Settlement, did not attract the Management and influence its behaviour. Indeed some of the acts and attitudes of the Management can be interpreted as designed to protect the independence of the Settlement as an organization and to enable it to

---

[1] *The Irrigation Bill, 1965,* published in Kenya Gazette Supplement No. 74 (Bills No. 20) Nairobi, 21 September 1965, became, with a few amendments *The Irrigation Act, 1966, No. 13 of 1966,* under which the National Irrigation Board was set up.

[2] *The Irrigation Act, 1966,* para. 24(1). However, para. 24(2) gave the Board, with the approval of the Minister, discretion to regulate the membership, powers and duties of the advisory committees.

[3] For a description and discussion of the organization and operation of the Settlement in 1964, see de Wilde *et al.,* op. cit., Vol. II, pp. 221-241.

survive in its environment. But the Management sought to free itself of interference from local officials and of controls from Nairobi not just for the sake of an independent command, but more significantly in order to be able to concentrate on its major objectives of increasing production and maximizing returns. In sympathy with these aims, the main criteria by which the Settlement was judged ceased to be human and political—the numbers of tenants settled and their morale—and instead became agricultural and economic— the statistics which could be presented to show production and returns. This is not to say that the tenants' interests were neglected. It is to say, however, that the dominant view was that what was good for production was good for the Scheme, and what was good for the Scheme was good for the tenants. Because of this intensifying attention to economic and productive aspects it is possible to speak of a core process, a process around which and in response to which activities and organization were oriented, a process which was perceived to present certain requirements which had to be met if it was to operate effectively. The core process on Mwea became the cycle of events and activities required by the growing, processing and marketing of irrigated rice. In determining the internal system of organization of the Settlement, technical knowledge of the core process now took the place of understanding of the tenants' behaviour and attitudes as the central consideration.

At first there was much ignorance. Uncertainty about what crops would grow, about pests, watering rates and markets, coupled with the speed at which the Scheme was implemented in the mid-1950s, spurred the Agricolas into investigating many crops.[1] On the black soils maize, beans, cotton and sugar were tried or contemplated at different times; but paddy rice proved the winner and trials for varieties, spacing and spraying were carried out. There were many imponderables: the threats that quelea might annihilate the crop, effectively countered by aerial spraying; the dangers of pest diseases, none of which in the event had proved really serious by 1967; ignorance about watering techniques; the problem whether to grow the rice during the short rains (November–December) or long rains (April–May), only settled in 1958 in favour of the short rains; the question of whether two crops of rice could be grown in a year, the subject of experiments for three years (1962 to 1964) before it was decided that the small additional yield did not justify the extra effort; and the issue of a second crop that might alternate with rice. As an alternating crop, fish[2] were considered and tried, but experi-

---

[1] Crop experiments are described in *Department of Agriculture Annual Reports* and M.I.S. Annual Reports 1957–1966.
[2] *Tilapia nigra*, indigenous to Kenya.

ments ended when several snags became evident, among them that the fish would be killed by the copper sulphate used to attack the snails which carried bilharzia. On the black soils annual monoculture of rice triumphed over all alternatives. On the red soils the position was more complex. Many proposals were put forward. The range of crops tried is a pleasure to recite, including as it does sunflowers, garlic, velvet beans, grapefruit, chillies, Chinese pumpkins, asparagus, hibiscus, cotton and tea, as well as plebeian onions, tomatoes and cabbages. The range of agricultural plans put forward was similarly extensive, including stall-feeding cattle with irrigated fodder, growing vegetables to sell to the tenants, and supplying canning factories with various crops. Experiments continued, and from 1960 onwards five tenants under special supervision were cultivating trial crops on small red soil holdings, though without conclusive results by 1965. The red soils attracted agriculturalists and inspired pipe dreams. Plans tended to embody the dominant agricultural ideal of the time: in the late 1950s and early 1960s the balanced mixed farming, with prominence given to manure, associated with the sturdy yeoman stereotype; in the mid-1960s, growing crops for canning for export; and in 1966, when co-operatives were acceptable, the co-operative cultivation by tenants of cotton on unirrigated red soils, which took place under Management supervision and made use of Settlement tractors. To the Management, however, the red soils appeared something of a side issue. Unirrigated and unsupervised as they were they could be cultivated by tenants and junior staff to supplement their subsistence, but they were not allowed to become a distraction from the black soils and the main chance—rice.

Curiously, in view of the importance of research to the economic success of the Scheme, the Research Station was at first part of a separate organization from the Management. The Research Officers worked not for the Manager, but for the Research Division of the Ministry of Agriculture. Physically, too, they were separate. Their quarters and the Research Station itself were at Wanguru, the other end of the Tebere section from the houses and headquarters of the Management.[1] They were also technically better qualified but less well paid than the first Manager. The Research Station's experiments were concerned more with water use than with yields, and until the early 1960s there was a marked lack of communication and liaison between the Research Station and the Management. Indeed, farming practices on the black soils evolved at first largely as a result of experience gained growing rice with the tenants, and later as a result

[1] See map on p. 133.

of experience brought in from British Guiana by the third Manager, rather than as a result of formal research.

During the process of consolidating and organizing the Settlement which gained momentum in 1960, the Research Station and staff were progressively drawn in under the Manager's control. It was agreed, first, that the Settlement was the customer of the Station and should say what research was wanted. Then in 1963 a committee, consisting of the P.A.O., the Agricultural Research Officer, Embu, and the Manager, was set up to co-ordinate research. Finally, when it appeared that research would have to be curtailed because of an economy drive, the Settlement took over financing the Station, which then came fully under the Manager. The Research Officer became an officer of the Settlement, and the Station was moved from Wanguru to a point, in the words of the Manager, 'nearer home',[1] that is, nearer to the Settlement Headquarters at Kimbimbi. While this process of incorporation was going on, investigations were concentrated on a narrower range of projects with clearer immediate economic value to the Settlement such as rice variety collection and fertilizer trials.

Research, experience from outside, and immediate experience on the Scheme contributed to a system of ideas about conditions and practices which would give the best economic results. These conditions and practices are here described as perceived imperatives, meaning that the Management perceived them as necessary for optimal economic achievement. The perceived imperatives did, of course, change as knowledge increased, and tended in the direction of securing better economic results. Since the Settlement provided a relatively stable habitat for rice, with adequate water and generally regular seasons, the perceived imperatives were themselves comparatively stable. In the early 1960s they included, for instance, sowing seed in late July and early August; preparing paddy fields to be flat, and with the surface puddled, that is, reduced to a wet, creamy consistency free of vegetation; careful transplanting of 4 to 5 week old seedlings on fields covered only with a film of water, but which should then be flooded to a depth of 4 to 6 inches, care being taken not to flood out the plants; maintaining water at that level, gently running, for the life of the crop; protecting the crop against birds, once the grain had begun to form; draining the fields before harvest, after which the rice was to be cut, threshed and bagged at a given moisture to minimize loss in storage; and synchronizing cultivation activities to reduce seepage and the possibility of a pest build-up. The value of these perceived imperatives was so

---

[1] Interview with the third Manager (1965). See map on p. 133.

firmly believed by the Management, and their achievement was so resolutely sought, that they can be considered almost genetic elements in the organization of the Settlement, which was developed and adapted to make their achievement more probable and more exact. This can be illustrated by the change from ox-drawn to tractor cultivation. Before 1960 all cultivation was carried out by oxen.[1] Fields were first dry ploughed and then, after flooding, puddled with a long plank called a levelling board. The ploughing was of a poor standard and damaged field levels. The levelling board, being tineless, did not eradicate any vegetation missed by the plough. The further disadvantages of this system have been summed up:

> The soil moving efficiency of the board was negligible. The cumbersome size of the tool militated against thorough work, particularly in the corners and along the sides of the fields. The strain on both oxen and attendants was very considerable. The Settlement provided the oxen, ploughs, yokes, chains and boards, the tenants operated them. A vast amount of land, manpower and paperwork was required to maintain and service the animals and equipment needed for this type of cultivation. Each tenant was responsible for the cultivation of his holding. It proved quite impossible to lay down and adhere to a cultural programme, due to the vagaries of the individuals on whom execution of the plan depended. It was not uncommon to have in a small area of 50 acres a spread in planting dates of nearly four months. Efficient water discipline and utilization were impossible, wear and tear on irrigation and drainage facilities was excessive, the possibility of a disease and pest build-up increased and harvesting was excessively and uneconomically drawn out.[2]

In late 1959 the Scheme had three herds each of one hundred work oxen for hire to the tenants, who were in effect acting largely like smallholder outgrowers for a marketing service which also provided the seed and the means of cultivation. It was in these respects a pre-industrial rural credit scheme, making use of capital development done by the Management and subject to some control through water supply, persuasion and sanctions. While it remained so it was difficult to approximate at all closely to the perceived imperatives of the crop, and the recorded deliveries of the tenants were low compared with those achieved later.[3]

Closer correspondence to some of the imperatives was sought

[1] See Plate 1.
[2] E. G. Giglioli, 'Recent Advances in Rice in Kenya—the Mwea Irrigation Settlement', *Agronomie Tropicale*, No. 8, August 1963, p. 829.
[3] For recorded delivery figures, see table on p. 135.

through mechanical cultivation which was introduced in 1960 on 1,575 acres, and in 1961 for the whole of the black soils of the Settlement.[1] Ploughing and puddling with oxen were abandoned. Instead a tractor-mounted rotavator, whose action itself helped to propel the tractor, was used to puddle the fields.[2] All vegetation was cut up and churned in, and by keeping the fields flooded until cultivation weed growth was prevented. Rotavation was started in April each year and completed in time for transplanting in August/ September. Productivity was improved by the better puddle and the complete absence of weeds at the time of transplanting. One result was that new fields now reached their full potential in their first year of cultivation instead of taking several years. Mechanization also made it possible to perform operations simultaneously throughout the Settlement at optimum periods. Whereas, before, transplanting and harvesting could sometimes be found in progress in contiguous fields, now all transplanting and all harvesting took place at the same, separate, periods. The danger of a pest build-up was reduced and seepage from flooded fields to dry fields, which had earlier been a problem, was largely eliminated. Economies were achieved in water use, which was of long-term importance in a country with more irrigable land than water, and control of water by the staff was simplified. There was also an improved opportunity for extension work. Within the mechanized cycle, tenants were constrained to carry out synchronically the successive operations of preparing nurseries, sowing seeds, tending nurseries, transplanting, weeding, scaring off quelea, harvesting, threshing and bagging. In contrast with what has been called the 'chaotic individualism'[3] of the ox cultivation system, Settlement staff could now call meetings before each operation, knowing that all tenants were about to perform it. Indeed, the programme of extension work made possible by mechanization was considered to be its most outstanding result.[4] Associated with these changes recorded deliveries rose to 24 bags per acre in 1960, and to 30, 32 and 34 respectively in 1961, 1962 and 1963, and the scatter of recorded deliveries was reduced.[5]

---

[1] The problems and practice of mechanical cultivation on the Settlement are described authoritatively in E. G. Giglioli, 'Mechanical Cultivation of Rice on the Mwea Irrigation Settlement', *East African Agricultural and Forestry Journal*, Vol. 30, No. 3, January 1965, pp. 177-181.

[2] See Plate 2.

[3] Giglioli, *Agronomie Tropicale*, op. cit., p. 828.

[4] Giglioli, *East African Agricultural and Forestry Journal*, op. cit., p. 180.

[5] 'Mwea Irrigation Settlement, An Analysis of Rice Yields, 1959 and 1960 Short Rain Crops' (mimeo), 'Mwea Irrigation Settlement, Analysis of Yields and Incomes for the 1961 Short Rains Rice Crop' (mimeo), and also for 1962 and 1963.

Mechanization generated its own problems and requirements, which affected the organization and work of staff and tenants. Strict water control was found to be necessary before rotavation to ensure that enough water, but not too much, was run on to the land at the right time.[1] Tenants were required to be present to move heavy but portable metal bridges at the time of rotavation to enable the tractors to move over the bunds between fields.[2] The efficiency of the Settlement became vulnerable, in a manner it had not been before, to staffing problems, since a drop in tractor maintenance standards and the resulting breakdowns could, as it did in 1965, threaten the continuation of the mechanical system.[3] Moreover, by increasing yields and synchronizing the harvest, mechanization compounded the difficulties of crop handling. The tenants reaped and threshed in their fields, and bagged the paddy which was then transported to reception centres, one of which was built for each of the sections of the Settlement. Reception, weighing, testing for moisture, drying to the correct moisture and rebagging were operations which required innovation and rationalization. The steady increase in efficiency is shown by these figures:

TABLE 6.2

Paddy handling performance

Year	Days of reception	Total receipts in bags	Mean number of bags received per day
1960	95	72,190	760
1961	140	138,347	988
1962	79	147,912	1,872
1963	64	149,704	2,341

Source: Manager's Report, March 1964

The imperatives perceived by the Management were not limited to the process of production within the Settlement; some were also derived from the feedback from the Settlement's environment. To take one example, complaints from the marketing organization about deliveries of overdry paddy drove the Management into

[1] Giglioli, *East African Agricultural and Forestry Journal*, op. cit., p. 179.
[2] See Plate 3.
[3] The inability to recruit maintenance staff led to such severe breakdowns that at one time some tenants were advised to return to ox cultivation. The situation was saved by a machine servicing contractor, and eventually by the recruitment of a Workshop Supervisor (M.I.S. Annual Report 1965, pp. 4-5).

attempting to speed up deliveries of harvested paddy to reception centres, and to exercise closer control over drying.

The organization and practice of staff and tenants that were developed and rationalized from 1960 onwards owed much to the groundwork of the earlier pioneering years. By 1960 there were procedures for handling all the main operations required in the Settlement, but these had usually been improvised under pressure. They were now streamlined and made more efficient, becoming more specialized and stable. At headquarters the filing system was reorganized and buildings were improved. In August 1960 senior staff duties were redefined and re-allocated. On the accounts side, the staff of four who handled 821 tenants' accounts in 1960 were heavily overworked in 1961 trying to deal with the increased number of 1,244. But following an Organization and Methods study[1] the system was simplified and in 1962 no difficulty was experienced in accounting.[2] In 1963 the Manager was able to express the formal specialization of the headquarters organization in a chart:

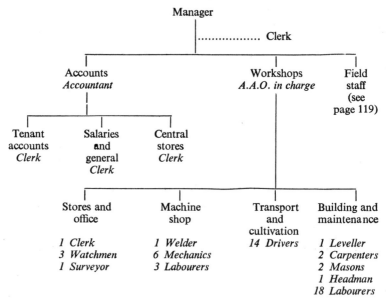

The Research Officer and Research Station were subsequently brought in under the Manager.

[1] 'A Review of the Organization and Administrative Activities of the Ministry of Agriculture, Animal Husbandry and Water Resources: Tenant Farmers Accounts, Mwea-Tebere Irrigation Scheme, January 1962' (mimeo).
[2] M.I.S. Annual Report 1962, p. 4.

Plate 1 Ox-cultivation, pre-1961. See page 115. Photograph by E. G. Giglioli

Plate 2 Tractor rotavation, 1966. See page 116

Plate 3 Tenants and tenant-employed labour moving metal bridges to enable tractors to cross bunds, 1966. The three tractors in the background are just finishing rotavation of the field. See pages 117 and 167

Plate 4 A small meeting of staff and tenants, 1966. From left to right: two tenants (standing); the Head Cultivator (leaning on stick); the Field Assistant; the A.A.O. in charge of the section; and tenants

Plate 5 Mwea: the visibility of deviation, 1966. An A.A.O. (nearer camera) and a Field Assistant looking at a section of drain which has not been cleared by the tenant responsible. See page 166

Plate 6 Mwea: a meeting of staff and tenants, 1966. The staff and a Head Cultivator are standing on the left. The tenants are sitting. At this meeting the A.A.O. warned those, including the tenant whose uncleared drain is shown in Plate 5, who had not performed the operations required by the cultivation routine

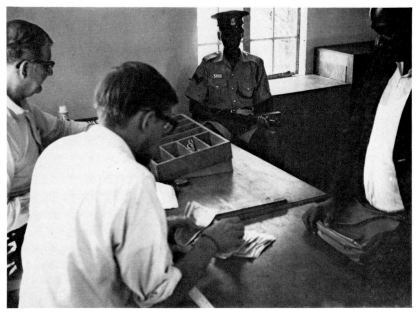

Plate 7 Mwea: the payout, 1966. From left to right: the Accountant; an American Peace Corps worker; a Kenya Police Corporal on guard; and the tenant who is being paid

Plate 8 Mwea: examining the pay slip after payout, 1966. On the left a tenant; on the right a Field Assistant. The pay slip gave details of deductions. See pages 184-185

Field organization, and with it the relationships of staff and tenants, were also gradually more closely defined and better understood. Each of the two sections of the Settlement, Tebere and Nguka, was the responsibility of an A.A.O. who supervised all cultivation activities in his section through two hierarchies, one of Field Assistants who were in charge of agricultural matters and who handled the tenants, and one of Water Guards whose formal responsibilities were for water control only. Expressed as a chart the organization of a section in 1966 was:

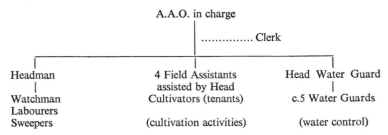

As with other formal organizations actual behaviour did not always accord with allocated roles. In particular, Water Guards commonly worked to Field Assistants, despite their nominal separation, in passing information and instructions on to tenants. The effective functioning of the system from the Management's point of view was, however, promoted by the criteria by which junior staff were judged: an absence of complaints and problems, reliable and accurate water control by the Water Guards, and high recorded deliveries from individual Field Assistants' areas. As long as these were met, informal deviations from formally allocated roles were not a matter of concern.

In securing change and efficiency, communications, both within the staff organization and between staff and tenants, were important. Contact and exchange between the Manager and junior staff took place through a revival of an 'African Staff Committee' which had met under the chairmanship of a member of the senior staff in 1959. The new committee had a syndicalist form of representation with members for different staff groups such as clerks and drivers. Under the Manager's chairmanship this committee met 14 times between July 1960 and December 1962 and discussed a wide variety of topics from the colour of uniforms to the opportunities for staff to obtain tenancies. In 1963 it ceased to meet when the Kenya Civil Servants' Union sought to take over the committee's function of negotiating with the Management on behalf of junior staff. By then, however, it had dealt with some of the main issues, such as junior staff housing, which slowly improved over the years. On matters of day-to-day agricultural management much of the communication among the staff took place informally in the field.

Communication between staff and tenants occurred in several ways. There were formal representative bodies—the Tenants' Advisory Committee at Settlement level and the Tenants' Liaison Councils at section level—which met once a month, providing an opportunity for the Management to put across its propaganda and instructions and for the tenants to make complaints and suggestions. Less formally, the A.A.O.s in charge of sections and their Field Assistants frequently met small groups of tenants about cultivation matters.[1] With mechanization, a regular series of public meetings held by Field Assistants was instituted so that they could indoctrinate the tenants about the next operation to be carried out in the cycle. In addition, as before, the supervision of cultivation took place in the field.

Some of the most important relationships and conventions were developed in those parts of the system where, at least at first, there was a conflict between managerial and tenant wishes. The Management was concerned with agricultural efficiency and high deliveries of tenants' rice. The tenants, however, had a wider range of social and economic interests. The Management sought to ensure the tenants' presence on the Settlement in order to perform the required agricultural operations; but tenants had social and economic interests and commitments off the Settlement which periodically demanded their absence. The Management was concerned with compliance with instructions on cultivation; but tenants for whatever reasons sometimes preferred not to carry these out. The Management, through the Water Guards, tried to regulate all water movement; but tenants sometimes had their own ideas about irrigation and moved water on their own initiative. The Management sought to maximize paddy deliveries to the Settlement organization, but the tenants often preferred to 'beat the system' and sell their paddy unofficially on a black market, or to retain more than was permitted for home consumption. To combat these problems the Management used sanctions and incentives. The sanctions of warnings, warning letters, prosecutions and ultimately eviction were employed especially to control water discipline, bad husbandry, poor maintenance of holdings and absenteeism. But the approach to managerial problems was psychological and economic (through incentives) as well as legal and disciplinary (through sanctions). For example, the black market in rice was attacked by larger and more prompt payments on the delivery of paddy to the Settlement. The greatest incentive for compliance

[1] See Plate 4.

by tenants to the requirements of the Management was the relationship between correct performance of operations and higher financial rewards resulting from higher yields. It was this, the strong common interest of staff and settlers, which maintained the Scheme system.

Through research and experiment, the reorganization of staff, the indoctrination of tenants and the manipulation of sanctions and incentives, the economic system of the Settlement was made more efficient. Many innovations, such as mechanized cultivation, the use of fertilizer, and individual instead of shared rice nurseries, were introduced. By the end of 1964 the Manager was able to report that the various operations were now looked upon as routine, 'every member of the staff knowing what to expect and what is required of him',[1] and that there was better harmony and understanding at meetings of the tenants' committees as the members had become 'sufficiently acquainted as to what matters lie within the realm of possibility as far as management is concerned'.[2] The Management had come to accept limitation by some of the tenants' wishes and attitudes, and the tenants had adapted to the requirements of the Management. That this was possible resulted largely from the success of the economic system which rewarded the tenants with money and food and the Management with prestige and other satisfactions. However, the extent to which the Settlement was able to prosper economically depended not only on internal efficiency but also on its relationships with organizations in its external environment and its adaptations to the opportunities and constraints which they presented.

The Settlement and other organizations

In the situation of greater managerial autonomy that prevailed from 1960 onwards, it might have been expected that the Management would pursue its earlier aims of forming a comprehensive administration on the Settlement and controlling 'all happenings'. The Management might have been expected, for instance, to attempt to acquire the magisterial powers and the power to select settlers which it had sought earlier. But both these ambitions were dropped. Prosecutions for irrigation offences were taken before the African Court at Wanguru, and when more settlers were required in 1963 the Manager was glad to leave the political function of selection to the Administration. Similarly, it might have been expected that just as the M.O.W. and the Administration had earlier pushed forward their boundaries of control by following the flow of a process, so now the Management would have attempted to follow through the production process

1 M.I.S. Annual Report 1964, p. 4.
2 Ibid., p. 7.

beyond the boundary at which the crop was handed over to a marketing agency. This had indeed been suggested by the General Manager who wanted the Scheme to market its own rice. But that was in 1959 at a time when the fight for roles was still important, and in any case the proposal was over-ruled by the J.I.C. and ceased to be an issue. Nor did the Management attempt to assert rights to perform new functions as they became necessary. Instead it sought organizations which could take them on. Thus credit was supplied by the marketing agency; banking facilities for tenants were provided by commercial banks; the transporting of paddy was put out to contract; the main road through the Settlement was maintained by the M.O.W.; health services were supplied by central and local government; and at one time even tractor maintenance was carried out by a commercial firm. Far from fighting for full control and self-sufficiency, the Management was now deliberately establishing links with outside organizations in order to obtain services for the Settlement.

The shift of objectives from comprehensive control to specialized and technical management can be illustrated by the epic of the aqua privies.[1] In 1957, £15,000 was made available, on an Administration vote, for piped water supplies to villages. At that stage there was a dispute between the M.O.W. and the Management, each wishing to be responsible for carrying out the work. The D.C. ruled in favour of the Management and some water supplies were installed. However the Provincial Medical Officer later argued that unless there was a very complete and expensive reticulation system, tenants would continue to draw water from the canals. His alternative suggestion, that most of the £12,500 remaining should be spent on sewage disposal by means of aqua privies, was accepted by the Local Committee and the Manager undertook installation. In 1959, 94 aqua privies were constructed, but the operation was not enthusiastically pursued in 1960. When the General Manager sought to put up an application to the Treasury for £6,600 to complete the aqua privies, the Manager saw the proposed programme as a threat to his more important economic activities. Not only would much of his time be taken up in supervision of scattered workers and contractors, but, he wrote: 'A rushed and expanded programme of aqua privy construction at this time will clash and compete very seriously with the vitally important existing work of improving drying and storage facilities at Tebere and Nguka.'[2]

[1] An aqua privy is a form of sanitation in which a cement tank with a soakage drain is supposed not to become full providing water is poured in regularly when the privy is in use.
[2] Letter, Manager to General Manager, 18 March 1960.

In 1960 only a few aqua privies were built, but in 1961 following alarm at what was believed to be a severe risk of inadequate sanitation a further 101 were completed.[1] Unfortunately, no department had developed a sense of ownership for the programme. Indeed, as soon as it was clear that the privies were not working as planned, avoidance action was taken and the Management was left alone reluctantly in charge. Approval had been given to build smaller aqua privies than originally designed, and the conservancy staff who were not closely supervised did not regularly pour in water as required. Consequently, the privies quickly became full. No exhauster could be hired, and the Settlement was forced to use its own staff and tanker for the task, to the disgust of the drivers, whose case was taken up by the Kenya Civil Servants' Union. The J.I.C. considered the problem and agreed with the Manager's suggestion that the County Council be invited to manage conservancy on an agency basis. But the problem was eventually buried in 1963 when the health authorities abandoned their previous insistence that pit latrines must be 30 feet deep, a depth impossible on the Settlement, and agreed to accept 15 feet, which was possible, as the new minimum depth. The Manager suggested that the tenants should dig their own pit latrines and all aqua privies should be sealed off, to the joy of future archaeologists. Where, as in this case, specialists could not be found, it was a relief to the Management who were amateurs in such matters to withdraw from responsibility.

This episode can be paralleled in changed attitudes of the Management to other non-productive activities including house construction, schools, water supplies, and village organization. In the atmosphere of *détente* from 1960 onwards, the Management felt less threatened by rivals, and with its secure legal foundation, its bureaucracy, and its undisputed authority in Settlement economic affairs, felt it could safely devote its resources of time and effort to the economic process. Without fear of take-over bids it could seek organizations in its environment which could provide specialized support and relieve it of distractions. Whereas in the organizational jungle of the later 1950s it had fought on all fronts, it now withdrew from many of them in order to concentrate attention on the core process of rice production.

In so concentrating, however, the Management encountered to a new degree of intensity problems arising from the parental controls of government. The installation of an outside telephone in June 1960 strengthened the tendency for the Manager to consult and work direct to the Ministry of Agriculture, and notably the Chief Agriculturalist in Nairobi. The Settlement became more of what was

[1] M.I.S. Annual Report 1961, p. 10.

known as a 'Nairobi baby'. But Nairobi was the seat of government regulations which often conflicted with the imperatives of the Settlement's economic system. Government regulations were designed to maximize accountability and minimize abuse of public funds, whereas for an agricultural estate run on industrial lines and required to maximize output and net returns, flexibility and adaptability were essential. The problems that arose and to which responses were made included financial accounting, the employment of contractors, stores accounting, and personnel management.

The government accounting system, while well equipped with cross-checks and skilfully arranged to provide auditors with means of assuring themselves that public money was not being used for private purposes, did not allow any commercially meaningful form of cost accounting. The government division of expenditure into non-recurrent and recurrent, as used in the Settlement's accounts, was not the same as the commercial division into capital and operating costs:[1] for instance, staff salaries were normally shown as recurrent expenditure even when the staff were engaged on what commercially would have been regarded as a capital improvement. Another difficulty was arriving at the costs of mechanical cultivation, which were required so that the charge to the tenants for that service could be estimated. Information on costs was widely scattered: in a fuel issue book, in motor vehicle log books, on three separate paysheets for three separate grades of staff, and in workshop records, so kept that a maintenance charge could not be arrived at. In addition, as government did not depreciate its vehicles, no depreciation charge could be derived from the accounts. In these circumstances cost accountancy of tractor operations was difficult. The problem was tackled by keeping two sets of accounts, one for government audit, and one for cost accounting. But even so there was no means for rapid and immediate assessment of cost performance so that the Management could tell what cost factors needed watching.[2]

Similar difficulties were experienced with regulations for employing contractors, according to which contracts could only be awarded after a lengthy set of procedures involving the D.C. and the Central Tender Board. When the Manager found that a crop transport contractor had inadequate lorries, he should, strictly according to the book, have spent weeks advertising for another, by

---

[1] Personal communication from Stephen Sandford.

[2] These and other problems of operating a would-be commercial concern within the ambit of government regulations were discussed in a paper by the Manager, 'Mwea Irrigation Settlement: Considerations on the Future', dated 24 April 1963. The discussion here also relies upon information provided by subsequent Managers and the Accountant.

which time the crop would have been lost. In the interests of the Settlement, he took the risk of appointing another contractor without delay. Again, when labour contractors were taken on to assist with crop handling at the time of harvest, the Manager needed to be able to hire or fire them at a moment's notice. The alternative, with an inefficient contractor, was to endanger the whole crop.

Stores accounting provides further examples. The procedures for writing off used stores were so cumbersome, requiring a Board of Survey to include the D.C., a member of the Settlement staff, and another officer, that it was more economical in staff time and the use of facilities to keep useless stores on the ledgers. A special boarding store was set aside for this purpose. One lorry, after the expiry of its working life, was kept on blocks for five years rather than divert staff effort to the unproductive task of securing its correct bureaucratic death certificate. A similar frustration was experienced over trade-ins and local sales. Local offers at £75 each for two boarded Jeeps had to be refused. The Jeeps were eventually auctioned in Nairobi in accordance with government regulations to fetch a total of £10.[1]

The most serious and difficult incompatibility between the Management's aims and government regulations concerned personnel policies. The government personnel system was designed to fit the clerical model, to provide stable and secure employment with regular hours of work in tasks in which routine presence on the job throughout the year was more important than intense or excellent work at any one period. With the ox-cultivation cycle, the fit between the clerical model and the actual work on the Scheme was not seriously out of line: since a wide scatter of activities took place concurrently throughout the year, junior staff roles were relatively undifferentiated, and the tenants did the muscular work. With mechanization, however, the requirements of the Settlement were those of a seasonal industrial process: high performance over long hours of work at certain periods of the year, with slack intervals between. For instance, during crop reception it was necessary to transfer staff from their normal occupations, but it was not permissible to pay them bonuses for overtime or for special short-term responsibilities.[2]

The most interesting and illuminating example is that of the tractor drivers, which demonstrates not only the conflict between aims and

---

[1] The Manager observed that 'The Manager of any private enterprise going in for a deal of that nature would be fired' (Ibid., p. 2.).

[2] In 1960, for instance, the three shift superintendents on the drier were paid Shs. 129, 176 and 437 per month for exactly the same work, their salaries being those carried by their normal jobs ('Memorandum on Improved Accounting and Managerial Flexibility', by the Manager, 9 September 1960).

rules, but also the growing sophistication of the Management's criteria and approach. An early rebuff to the Management's attempts to provide rewards and incentives came in 1959 when an application for permission to pay drivers overtime failed. Unable to pay overtime, the Management resorted to a reward that was invisible in the government system: granting extra leave in slack periods. At first the drivers worked well. In 1960, the Manager wrote: 'Great credit is due to the tractor drivers engaged on mechanical cultivation. These men have been working a 12 hour day, 7 days a week for nearly three months cheerfully and conscientiously under most unpleasant physical conditions.'[1]

But this did not last. The system of incentives through leave was said to have a 'never-never touch' and not to be understood by the drivers.[2] Further, it was difficult to dismiss an inefficient driver. The Manager put it that:

At present a tractor operator who is merely incapable but otherwise obedient, willing and guiltless of any major crimes is unfireable for practical purposes; this means that the tractor crawls along for the entire cultivation producing about half as much work as it should.[3]

A campaign, at first unsuccessful, was mounted to persuade the Treasury to allow a bonus system. In the absence of such a system, drivers' morale declined, and ageing equipment liable to breakdowns contributed to the drop in daily performance from 5·04 acres prepared per tractor day in 1962 to 3·84 in 1964.[4] Only towards the end of the 1964 rotavation programme did the Treasury finally agree to bonuses, and the agreement was to a system more limited than had been requested. There was an improvement in performance as a result, but the Management was beginning to take a more inclusive view of the process, and to be concerned more with economic and long-term aspects of tractor use than with daily performance. The incentive system based on acres per day did not encourage good maintenance and there were frequent breakdowns. In 1966, therefore, the hours of work were limited and bonus payments were made to drivers on monthly performance, thus placing a value on steady work without breakdowns.

Throughout the period 1960–1966, the Management was frustrated by government regulations such as those described. Although there was progressive relaxation of some of the rules, the Manage-

---

[1] Manager's Monthly Report for November 1960.
[2] 'Memorandum on Improved Accounting and Managerial Flexibility', op. cit.
[3] Ibid.
[4] Manager's Quarterly Report, April–June 1964.

ment were placed in a series of embarrassing dilemmas in which action required in the interests of the economy of the Settlement and of the tenants conflicted with the formal rules of government. The Management was driven to trying ingenious solutions, one of which was to seek, or to make a show of seeking, a new and more permissive parent. Two interesting proposals resulted, in 1960 and 1963 respectively, and concerned the two main organizations with which the Settlement was establishing a close and reciprocal relationship—the marketing body and the local government.

In order to market its rice the Settlement was faced with a choice of methods: sale by the Scheme, which though sought by the General Manager was rejected by the J.I.C.; or sale to a governmental or para-governmental organization. In the event, the Settlement's rice was always marketed through a government organization or board.[1] The successive Marketing Boards provided the Settlement with valuable services: they bought the whole rice crop at a guaranteed price, received and stored it at railhead, sold and distributed it to millers dispersed over Kenya, and allowed the Settlement credit to enable the Management to pay out the tenants soon after the crop had been harvested and delivered.

The frustrations of operation within the confines of government regulations, coupled with the close relationship that developed between the Settlement and the Marketing Board and the latter's relative autonomy in accounting and personnel matters, sparked off the first proposal. In 1960 the Marketing Board was unofficially approached with a view to its taking over most of the Settlement's accounts, a measure which it was thought would allow a greater degree of flexibility; for instance in paying overtime to staff. The proposal was put up by the P.A.O., but although sympathetically received in the Ministry of Agriculture it died in central government. This suggestion of elopement may, however, have been one of the factors which persuaded the government parent in the years that followed to adopt at least some relaxation of its rules in relation to the Settlement.

A similar initiative was considered in 1963, at a time when it appeared that irrigation would become a regional and not a central government responsibility. At the same time, co-operatives were fashionable and there appeared a danger that there might be pressure for the Settlement's organization to be modified in a co-operative

---

[1] These were: first, the Maize and Produce Control Organization, until 31 July 1959; then the Central Province Marketing Board, until 1 September 1964; and finally the Kenya Agricultural Produce Marketing Board. For a discussion of Provincial Marketing Boards see *Report of the Committee on the Organization of Agriculture*, Government Printer, Nairobi, 1960, pp. 75–82.

direction. Agricolas were, however, agreed that co-operative control would be disastrous for production. The expected transfer to regional government and the co-operative threat stimulated the Manager into considering the future of the Settlement. He argued that the Settlement needed a high level of control and discipline, the flexibility of commercial accounting, an enlightened personnel policy, and the ruthlessness of private enterprise. The only local body able to supply these was the A.D.C. which had adequate administrative capacity, a system of commercial accounting and, with its annual income of the order of £13,000 from the rice cess, a strong interest in the Settlement.[1] But whereas the earlier move towards the Marketing Board was an attempt to avoid the government parent, the contemplation of grafting the Settlement on to the A.D.C. arose only when it appeared with regionalism that the parent might abandon its child. As soon as it became clear that regional government was only a passing phase, the idea was quietly forgotten.

From 1964 onwards, as the authority of the central government was reasserted over the regions, preparations went ahead for the creation of a National Irrigation Board. When, on 1 July 1966, the Settlement's administration was transferred to the Board, it had at last found an institutional framework of methods and rules appropriate to the pursuit of the economic aims of a commercial undertaking. The relationship of the Settlement to the Board, however, was different historically from those contemplated earlier with the Marketing Board or the A.D.C. For the National Irrigation Board, though now a new parental and controlling organization, had itself largely been generated by the experience of Mwea, for which its constitution was especially adapted. The Settlement organization and the individuals involved in it, instead of adopting a new parent, had largely created their own.

Further growth

In the early 1960s the Settlement was no longer judged by political criteria. Its colonization at the tenant level by the local people removed its rationale as a place to settle landless Kikuyu, and the progressive exclusion of the Administration and the greater dominance of professional agriculturalists in the Management permitted and encouraged the shift towards economic attitudes. The very stage of growth of Mwea, with production and organization as the principal activities following settlement, also directed attention away from social and human considerations and towards economic and agricultural aspects of organization.

[1] 'Mwea Irrigation Settlement: Considerations on the Future', op. cit., p. 6.

Economically, Mwea was pronounced a success. At the end of 1960 it was said to have 'proved itself as an economic undertaking',[1] and at the end of 1961 it was described as 'fully consolidated and an economic success'.[2] While the other two major irrigation schemes, Perkerra and Tana, were making a loss, the surplus of revenue over recurrent expenditure for Mwea for the financial year 1960/61 was calculated as £26,000. In 1963 one authority even wrote that Mwea made 'a handsome profit more than covering the net loss on the other two (irrigation schemes) and quite adequate to repay capital development costs'.[3] In fact, enthusiasm about the economy of the Settlement and a rosy view of its viability were encouraged by a continuation of some of the permissive and protective non-economic attitudes of the earlier political phase. This can be illustrated by the issue of capital repayment. In accordance with the original intention that the capital cost should be repaid out of Settlement revenue, the question of repayment was raised in 1961. From the point of view of the Settlement as an organization, a nice balance had to be struck: on the one hand, it was desirable to be regarded as economically successful; on the other it was not desirable to be so successful that repayment of capital could be expected from revenue surpluses. Thus, forward estimates of revenue and expenditure for 1961–1966 prepared by the Manager in 1961 showed only small net annual surpluses. In the light of these the Settlement could continue to be regarded as an economic success, but they were small enough to justify quietly dropping the question of capital repayment. Had the principles of commercial accounting which were desired to make the operation of the Settlement more economic also been applied to the capital cost, it would not have appeared so successful.[4] It was always possible, however, to argue back to the political origins of the Scheme and to state that in the early stages expenditure had been incurred and risks taken which no commercial undertaking would have countenanced. To saddle the Settlement with the burdens of depreciation and capital repayment would therefore be unjust. Its earlier political purpose was,

[1] *Department of Agriculture Annual Report 1960*, p. 50.
[2] *Department of Agriculture Annual Report 1961*, p. 67.
[3] *A National Cash Crops Policy for Kenya*, Government Printer, Kenya, 1963, p. 44.
[4] In a summary of the financial performance of the Settlement made at the time of the take-over by the National Irrigation Board in 1966 the capital cost from the inception of the Scheme to 30 June 1966 was estimated as £787,368. The same summary showed for the six financial years 1960/61–1965/66, a total operating surplus of revenue over recurrent expenditure and depreciation of £15,345 ('National Irrigation Schemes: Consolidated Capital, Expenditure, and Revenue Figures for the 11 Years to 30 June 1966 (amended)' (mimeo), 28 May 1966). No attempt is made here to carry out an economic evaluation of the Settlement which will be the subject of a forthcoming study by Stephen Sandford.

thus, an asset to the Settlement, since it could be used protectively to avoid external drains on its resources, and to buttress the image of economic success.

The question of the economic performance of the Settlement is complicated by the marketing structure. The Marketing Board paid a guaranteed price to the Settlement, and built up a financial reserve out of the surplus between its receipts from sale of paddy on the one hand, and its charges for costs and payments to the Settlement on the other. Up to 1966 this surplus had never been used, and had accumulated £257,000.[1] The commercial viability of the Settlement varies enormously according to whether this sum is regarded as inside or outside its economic system. In practice, it was regarded as outside. However, the existence of this reserve was a powerful form of assurance for the Settlement organization that it would be able to maintain the returns to the tenants, and thus sustain itself and the Settlement's economy.

Whether the Settlement could expand was closely related to its economic position. This in turn depended not only upon production but also upon the market environment. One reason why expansion was halted in 1960 was uncertainty of an adequate market for rice from a larger acreage. With the inefficient and dispersed milling arrangements in Kenya there was no prospect of Mwea rice competing on the world market without a heavy export subsidy. In the local market to which it was therefore confined the major imponderable was whether the Asian community, the main consumer, could adapt its tastes to the slightly glutinous nature of Mwea rice, and whether the potentially very large African market would develop. In the event, the market posed no serious problems and after 1960 fears of a limited market did not restrain plans for expansion.[2]

The limiting factors were, rather, finance, and to a lesser extent land. Financially the Settlement was thrown back on its own resources when construction was completed in mid-1960. Any works or improvements had to be financed from the Settlement's own revenue, which was mainly derived from the water rates charged to tenants and deducted from the proceeds of their crops. Government policy in 1960 was cautious. In the words of the development programme for 1960–1963, it was intended: 'to review the present halt in expansion in two or three years' time in the light of available finance, of further suitable tenants, and of the demand for irrigable

[1] Interview, Officer-in-Charge, National Irrigation (1966).
[2] It remains to be seen at the time of writing (April 1968) whether the market will be seriously affected by the Asian exodus of the second half of 1966 and early 1967.

cash crops, to see whether further expansion would be justified up to the theoretical optimum level of 13,580 acres'.[1]

Not being required to repay the capital invested, the Settlement quickly acquired a small surplus. The Local Committee discussed the uses to which it might be put, and a strong divergence of views was noted. The medical representative championed a pure water supply; others favoured other forms of internal development to benefit the tenants. But the Manager's proposal prevailed when it was agreed to spend the surplus on a small extension. In the latter part of 1962 a small development programme started on 320 acres in the Nguka section. As the original organization had been disbanded a new one, which proved cheap and effective, was set up. Local contractors were employed on the structures, prospective tenants on the field work, and prison labour on the drains and furrows.[2] A slightly larger area was developed than originally intended and 94 new tenants were selected through the Administration and settled. The popularity and credit-worthiness of the Settlement were exploited by financing the tenants' house-building through bank loans. In 1963, in addition to this extension, tenants themselves reclaimed 126 acres in small fields in the Nguka section, and a further small extension was started in Tebere near Wanguru.[3] All these small additions were within what might be described as the natural boundaries of the existing Scheme, all north of the Nairobi – Wanguru road. They were small pockets of land which had been left out of the original development programme and which were now ingested and integrated into the irrigation system.

In the meantime preliminary action had been taken for a larger extension on the Nguka side. It was to be outside the existing irrigated areas but to use the Nguka headworks and canals. Development was much more methodical than in the early days of the Scheme. In 1961 water availability was assessed: the conclusion, that 4,000 acres more could be irrigated on the Nguka side, was favourable.[4] The market for rice was not considered a constraint, although a World Bank Mission urged caution.[5] The M.O.W. carried out topographical surveys and planned the field layouts. Soils were examined. Estimates of revenue and expenditure were drawn up for a scheme expanded by a first phase of 2,000 acres to a total of 7,000 acres, and by a

---

[1] *The Development Programme 1960/63*, Sessional Paper No. 4 of 1959/60, Government Printer, Nairobi, 1960, p. 29.

[2] M.I.S. Annual Report 1962, p. 4.

[3] M.I.S. Annual Report 1963, p. 4.

[4] 'Notes on Water Availability and its Influence on Irrigation Development Proposals', by the Senior Irrigation Engineer, 5 June 1961.

[5] *The Economic Development of Kenya: Report of an Economic Survey Mission*, Government Printer, Kenya, 1962, p. 56.

second, subsequent phase of 5,000 acres to a total of 12,000 acres. An application to the United Kingdom Committee of the Freedom from Hunger Campaign requesting £163,000 for the first phase was successful, and the first instalment was available by July 1963. Moreover an organization for the preparation of new fields had been built up during the work on the small additions, and this organization was available to carry out the extension.[1]

The extension required land over which clan elders tried to maintain their claims. The proposal to extend the Nguka section to the south-west bogged down, despite repeated efforts by the D.C., over the issue of subsequent control of the irrigated land. In November 1962 the clan committee agreed to rice expansion but on condition that for some areas at least the land would only be leased to the Management, with the implication that it had not been set aside. The committee also wanted a say in the allocation of tenancies, wished to decide on registration for individual title of land not required for rice development, and urged the Government to increase local responsibility in the management of the Settlement. An attempt to reconcile the Agricolas' insistence on autonomy of management with the clans' claims was made in March 1963 when registration of land in a co-operative was suggested, with individuals' shares proportionate to their land holdings. It was thought that eviction by the Management might then be possible. But, for all the ingenuity of this suggestion, the J.I.C. took it that the clans wished to maintain some degree of control after irrigation development, and rejected this as utterly unacceptable. Nor did the desire of the clan elders to reacquire the red soils within the Settlement, which were being cultivated by tenants and junior staff, have any success. The elders were weakened by opposition from the younger generations and also lacked the support of Members of Parliament. Against the resolute 'foreign' organization which, in their view, had taken over their land, the elders were virtually powerless. The effect of their agitation was merely to divert development from the most controversial areas to the south-west to the land to the south over which there was less contention.

Work began on the extension in 1964, and the Settlement gained a steady momentum of growth in terms of acres cropped and tenants settled.[2] The 2,000-acre Freedom from Hunger extension was completed in 1967, and in November 1967 an agreement was signed between the Kenya and West German Governments for a £400,000 development loan to finance a further 3,000-acre extension to bring

---

[1] The organization, which involved the M.O.W., Settlement staff, tenants and small local contractors, is described in M.I.S. Annual Report 1963, p. 14.
[2] See Table 6.3 on p. 135.

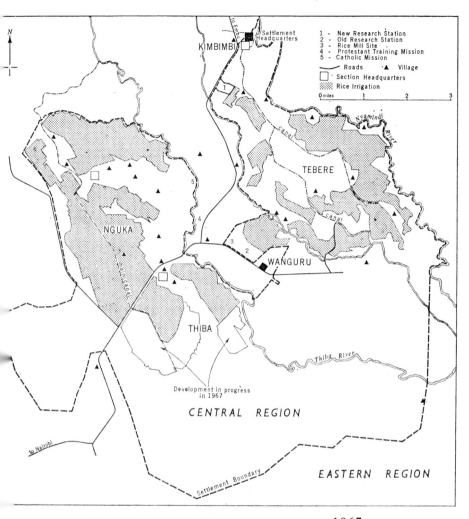

THE MWEA IRRIGATION SETTLEMENT IN 1967

the total to 10,000 acres.[1] By this time the processes of expansion had become well understood, and growth itself was becoming routine.

Expansion was not limited to physical extension. There were continuing attempts through fertilizers to increase productivity. There were also signs that the earlier specialization on rice growing was being tempered by an interest on the part of both tenants and

[1] *East African Standard*, 1 December 1967.

Management in diversification. Experiments continued on the red soils. More immediately important, however, was cultivation of cotton on the unirrigated red soils, using Settlement tractors, aerial spraying for pests, and a co-operative form of organization among the tenants. In 1966 131 acres yielded cotton with a gross value of £2,789, and following these encouraging results the experiment was repeated in 1966/67.

Apart from physical expansion the most extensive field for potential colonization by the Settlement was along the rice process itself. At the input end there was no scope, since rice seed was grown on the Settlement. But at the output end, when the paddy crossed the boundaries of the Scheme, there was an opportunity. Considerable diseconomies were involved in the system of shipping unhusked rice to mills dispersed over Kenya, not merely because of the distance which the husks had to be transported, but also because the mills were working below capacity and therefore being forced to maintain high charges for milling. In 1963 a Rice Processing Committee was convened to examine the proposal that a rice milling factory should be set up on the Settlement. This was followed in 1964 by the formation of a Tenants' Thrift Society with support and assistance from the Management, with the aim of raising part of the capital required for a mill. The Society's formation can be interpreted in various ways: as a defensive response by Management to the potential threat of a co-operative take-over of the Settlement; as an expression of the growing sense of economic community of interest among the tenants; and as a means, recognized by both Management and tenants, whereby together they might achieve an extension of the Scheme along the economic process to their mutual benefit, the Society providing political acceptability and the Management technical and financial expertise. By the end of 1966 the Society, now the Mwea Co-operative Credit Union, had accumulated £35,000 subscribed by its members, and machinery had been bought and a manager recruited for a mill to be established on the Settlement and jointly owned by the Credit Union and the National Irrigation Board.[1] The Settlement was on the way to becoming a processing and possibly marketing as well as a producing organization, and the tenants' leaders were moving closer to managerial positions.

Expansion was not, however, limited to developing new land, settling new tenants, and extending controls along the economic process, for the Settlement was beginning to influence the patterns of development and organization elsewhere. Within Kenya, as we have seen, the creation of the National Irrigation Board was partly a response to the needs of Mwea, and its form was partly based on the

[1] M.I.S. Annual Report 1966, p. 5.

TABLE 6.3

The expansion and recorded production of the Mwea Irrigation Settlement, 1958–1967
(Rice system only)

	A Black soils harvested (Acres)	B Total tenants	C Paddy (rebagged) (Bags[1])	D Mean recorded delivery[2] (Bags per acre)	E Mean net income per tenant[3] (Shillings)
1958	565	372	12,330	21·82[4]	Not available
1959	1,231	824	23,680	(19·24)[5] (15·78)[5]	(1,912)[5]
1960	3,281	1,244	79,760	24·31	2,606
1961	4,958	1,246	149,998	30·25	2,851
1962	4,973	1,246	159,812	32·14	2,721
1963	4,965	1,340	167,474	33·73	2,812
1964	5,465	1,340	165,864	30·99	2,483
1965	5,541	1,484	156,711	28·28	2,322
1966	6,407	1,588	199,162	31·08	2,549
1967	6,981	1,735[6]	190,889	27·21	2,371

Principal Sources: Mwea-Tebere Irrigation Scheme Annual Reports for 1958 and 1959; M.I.S. Annual Reports for 1960 to 1966 inclusive; 'Mwea Irrigation Settlement: An Analysis of Rice Yields—1959 and 1960 Short Rain Crops'; 'Mwea Irrigation Settlement: An Analysis of Yields and Incomes for the 1961 Short Rains Rice Crop', and also for 1962, 1963, 1964, 1965, 1966, and 1967.

Notes:
[1] A bag is 160 lbs. of clean, well-winnowed paddy at 13½% moisture.
[2] The mean recorded delivery does not include rice retained by tenants for their own consumption or for sale on the black market. Estimates for this varied considerably, but the order of magnitude was probably 10 bags per tenant per annum.
[3] The net income is taken as the payment to the tenant for rice delivered, after the deduction of Settlement charges. The cost to the tenant of any labour employed is assumed in the sources quoted (the Analyses of Yields and Incomes) to be equivalent to the value of rice retained.
[4] The Annual Report for 1958 suggested that the actual yield might be higher than this, since some of the 'acre' fields were less than an acre. The relatively high yield for this early stage of development may be partly explained by the fact that of the 565 acres harvested, 209 were under direct cultivation and therefore not subject to rice retention by tenants.
[5] 19·24 bags per acre is derived from the 1959 Annual Report, while 15·78 is given in 'An Analysis of Rice Yields for 1959 and 1960, Short Rain Crops'. The latter source gives detailed breakdowns of yields by subdivisions of the Scheme, but does not give acreages or deliveries. It is not clear which figure is to be preferred. The mean net income per tenant of Shs. 1,912 is based upon the mean yield of 15·78 bags per acre.
[6] The additional tenants were settled in May 1967 and did not cultivate for the 1967 harvest. The mean net income is calculated on the basis of the 1,589 tenants who actually cultivated the 1967 crop.

experience gained there. But beyond this the existence of the Board made easier the transfer of personnel and practices from Mwea. In 1967 the Settlement became a training ground for junior staff for other irrigation schemes, and at the senior staff level one of the A.A.O.s was posted to become Manager of the Tana Scheme. Moreover, when designs were made for the 2,000-acre Ahero Pilot Project in Nyanza Province, the Mwea pattern of one-acre fields to be cultivated by tenants in four-acre units was reproduced, and it seemed likely that the Project, to be based primarily on irrigated rice, would follow not only the physical but also the legal, organizational and economic system already established on Mwea.

In 1967 it was not only within Kenya that Mwea was becoming influential. Its strategic situation near the international airport of Nairobi, its convenience for a day trip from that city, and the growing numbers of political VIPs, nomadic experts and (it must be admitted) research students who were in circulation, combined to drawn an increasing flow of interested visitors to the Settlement. The benefits of such visits are not calculable, but it would be surprising if irrigation developments in other countries in Africa, and perhaps even outside Africa, were not affected. In the field of complex, disciplined settlement projects, it does not seem an exaggeration to suggest the Mwea model of organization was beginning to supplement the Gezira and *moshav* models which had been dominant in Africa earlier.

Part III

Settlement Schemes as Organizations

Settlement schemes can be examined from various points of view, among them social as communities, economic as production units, and political as interest groups. None of these approaches could, however, hope to be watertight; each requires the others if the whole is to be understood. An attempt is now made to examine settlement schemes in terms which incorporate the more significant elements of these approaches by considering them as entities in which people, activities and processes are interdependent and systematically related to one another: in short, as organizations. In Chapter 7 some of the characteristics and initial attitudes of the actors in scheme situations—the staff and settlers—are examined. In Chapter 8 the analysis moves on to their evolving aims, interactions and adaptations within scheme situations. In Chapter 9 settlement schemes are considered as organizations which adapt to their environments, and survive or fail to survive within them.

137

The Actors: Staff and Settlers

As on Mwea, so on other settlement schemes in anglophone Africa about which there is relevant information, there has usually been a distinction between staff and settlers. The differences can be identified along many dimensions, among them roles, housing, dress, terms of service, and the attributes brought to schemes. In terms of roles staff have been the supervisors and controllers, with some degree of responsibility to an outside or superior agency, while settlers have been those over whom supervision and control have been exercised. Staff have usually lived in housing of a different type and better standard than settlers, set apart in separate residential areas. Staff have often been distinguished from settlers in dress, most often some form of khaki uniform, or overalls in the case of drivers and artisans. Staff have also tended to lead a separate social life from settlers. Perhaps the most important distinction, however, concerns what may broadly be called 'terms of service'. Typically, staff, being almost always employed by government or by a government corporation, have had regular salary payments, often with permanent and pensionable terms, and have been subject to transfers. Settlers, in contrast, have received seasonal payments dependent on several factors, including the success or otherwise of a crop, and it has usually been intended that they should become 'settled' through the development of a permanent relationship with a particular piece of land. In part these various differences have been associated with the nature of evolving scheme situations. In part, and prior in time, they have been associated with the characteristics and attitudes which staff and settlers have presented to schemes, and especially with their ethnic, cultural and occupational attributes. When these characteristics and attitudes have been examined, as they will be in this chapter, it will be easier to interpret the manner in which organizations, of which these people form a part, develop on continuing schemes.

In some respects staff fall naturally into two groups: senior or managerial staff; and junior staff. This division has corresponded not only with responsibility, roles and housing, but also frequently with race. The reservation of senior posts for Europeans was abolished in West Africa in 1946, East Africa in 1954, and Central Africa in 1960,[1] but in practice the Africanization of senior posts in settlement schemes has been slow. In most settlement schemes for most of the time span of their existence, the senior posts have been held by Europeans and the junior posts by Africans. On Mwea this identity of post and race was recognized in the 'European Staff Meeting' which met five times in 1960 before reappearing as a 'Senior Staff Meeting' while junior staff met the Manager in an 'African Staff Committee' until the end of 1962, after which its functions were partly taken over by the Kenya Civil Servants' Union. This correlation of seniority with Europeanness and subordination with Africanness is an important factor for most schemes in understanding relations between senior and junior staff, and between senior staff and settlers.

Some characteristics of senior staff

A serious difficulty in generalizing about senior staff is the personality variable. For whatever reasons, Europeans who sought out colonial situations were often eccentric to some degree, and peculiar bents and enthusiams were allowed free rein in the isolated and autonomous commands afforded by many settlement schemes. Where policy was determined largely at a local level, as with projects initiated by the Administration, personal interest and whim could strongly influence the philosophy and style of a scheme. The range of attitudes possible can be illustrated by the contrast between, on the one hand, the D.C. responsible for the Zande Scheme in southern Sudan who had a paternal concern to protect and control the Azande and to save them from the evils of village or urban living,[2] and, on the other, the D.O. in charge of Shendam in Nigeria who was a strong advocate of self-help to enable the settlers to stand on their own feet[3] and who placed them in villages. An example of a settlement scheme allowing the satisfaction of a personal enthusiasm is provided by South Busoga where one of the D.O.s in charge has been described as having 'a passion for D.8 bulldozers',[4] as a result of which roads were constructed to the detriment of the compact settlement required to prevent the spread of sleeping sickness. Even where policy was determined by a parent organization, there was still considerable scope for individual variations where settlements were scattered over a wide area, as with the eight Pilot Village Settlements in Tanzania,

[1] Adu, op. cit., p. 18.
[2] Reining, op. cit., pp. 133 and 135–136.
[3] Hunt, op. cit., p. 16.
[4] Watts, op. cit., p. 144.

though less where a bureaucracy was fitted to a continuous territorial unit as with Gezira and Managil in Sudan, and many of the Kenya Million-Acre Settlements.

Recruitment of expatriates was influenced by the permanence and promotion prospects of work as a settlement manager as perceived by potential recruits. In this Gezira stands alone. In its earlier years it could offer a secure career structure to a manager, who might start as a block inspector, rise to become a group inspector and finally, as Arthur Gaitskell did, become Manager for the whole scheme.[1] The very size of the Gezira senior staff, which numbered about one hundred, must have encouraged a sense of permanence and possibility of promotion. By delaying Sudanization of Gezira, an attempt was made to maintain among British staff a feeling that they had a future on the Scheme, but the great majority left in a short period when Sudanization came, perhaps for the very reason that they had been led to expect a long-term career and were disillusioned when this expectation was shown to be false.[2] Other settlement schemes never appeared to offer stable or lasting employment. There were five main inter-related reasons for this. First, schemes had a high failure rate. Second, the terms of service offered were often short-term contracts, which might or might not be renewed. Third, many settlement schemes were intended to have only a limited period of management before being handed over either to normal administration, as with the tsetse resettlements, or to a combination of normal administration and settler control, as with the Tanzanian Pilot Village Settlements and the Kenya Million-Acre Settlements. Fourth, particularly in the later 1950s and the 1960s, it did not seem that European management of settlement schemes would last long in the face of pressures for Africanization. Fifth, prospects of promotion were generally poor. For example, promotion for an irrigation scheme manager in Kenya before the setting up of the National Irrigation Board could only be achieved by moving out of irrigation and back into the normal structure of the Department of Agriculture. An irrigation manager could easily feel he had been forgotten, that he had been posted to a backwater, and that he was being overtaken by his peers in the Department. It is true that a few expatriates gained rapid temporary promotion during Africanization, but they were exceptional. Most settlement managers neither achieved nor expected to achieve direct advancement in their posts.

In consequence, it was necessary for recruiting agencies to offer special inducements or to be flexible in the standard or type of person they would accept. Some of the staff who were recruited to implement

[1] Gaitskell, op. cit., p. 99.
[2] For a discussion of Sudanization and British staff see ibid., pp. 321-331.

the Kenya Million-Acre Scheme were probably attracted by the prospect of a higher salary and greater responsibility in the short term than they had in their current posts in the Department of Agriculture. In other cases staff were accepted who lacked formal training or relevant experience, and whose Europeanness was sometimes their most important qualification. In southern Rhodesia some of the older European Land Development Officers in charge of the Sabi Irrigation Projects were appointed before any training or experience was required.[1] In Nigeria it was recognized that the Development Officers responsible for resettlements could not be expected to provide all the technical knowledge required of them.[2] The amateur approach of some ALDEV staff in Kenya has already been discussed.[3] Even in Gezira recruitment of British staff was 'far more concerned with character than technical qualifications'.[4] Sometimes the men recruited to run schemes were 'bush' characters who 'understood the African'; and indeed, a little to the surprise of European liberals, they sometimes achieved very effective working relationships with African staff and settlers. In other cases recruitment outside Africa brought in people who were unable to understand their new working environment. One of the many shortcomings contributing to the failure of the Niger Agricultural Project was the fact that none of the senior staff had ever worked in Nigeria before.[5] Many of the managers of Uganda Group Farms were young graduates of British agricultural colleges with a technical competence in mechanized agriculture, but ignorant of the customs and culture of the people whom they were to manage.[6] There was a general tendency for European senior staff to be either locally recruited, in which case they had local experience but rarely formal qualifications, or recruited abroad, in which case they might have formal qualifications, but rarely local experience.

The flexible attitude to recruiting expatriates was reflected in a readiness to tap without much discrimination a variety of reservoirs of talent. With some simplification, three clusters of men can be identified: colonial wanderers; the young and the idealists; and international experts. The colonial wanderers, a heterogeneous group of versatile men, were first on the scene. Typically, they had experience in several British colonies. Their numbers were swelled after the

[1] Roder, op. cit., pp. 171–172.
[2] Colonial Service Second Devonshire Course, Cambridge 1952–53, 'Rural and Urban Settlement in the Colonies: Report of the Study Group on West Africa. (mimeo), p. 13.
[3] See above, pp. 52–54.
[4] Gaitskell, op. cit., p. 325.
[5] Baldwin, op. cit., p. 194.
[6] Charsley, 'The Group Farm Scheme in Uganda: a Case Study from Bunyoro', (typescript), 1967, p. 16.

Second World War by restless ex-servicemen, some of whom stayed on after demobilization in East Africa, while others went from Britain to Tanganyika to work on the Groundnut Scheme and remained after its demise.[1] Latterly, many of these colonial wanderers responded to national independence movements with a southward drift through the continent, starting with staff from Gezira in the mid-1950s and later including others from East Africa, numbers of whom migrated to Central Africa, the former High Commission Territories, and South Africa. The young and the idealists, who partly replaced them, had become common by the middle 1960s, and included agricultural graduates and diploma holders, American Peace Corps workers (some of whom assisted with the Kenya Million-Acre Scheme) and participants in the British Volunteer Programme. The third group, the international experts, often associated with the Food and Agriculture Organization of the United Nations, became a source of staff in most countries after independence. They were a most diverse group, as unpredictable as they were international.

The recruitment of African senior staff was equally pragmatic and permissive. Not only in Kenya, but in all countries except Rhodesia, as expatriates left and as new organizations were set up to carry out the programmes associated with independence, Africans were faced with wide opportunities for promotion and better jobs. Settlement work, with its physical isolation and often limited prospects, was among the less attractive careers, if indeed it could offer a career at all. Even on Gezira where there was a career structure recruits tended to be those who, for one reason or another, were not well placed for promotion or who felt frustrated.[2] Elsewhere settlement agencies struggling to Africanize offered employment to local people whose careers had gone wrong, whether through injudicious political alignment, dishonesty, or bad luck, and who found it difficult to obtain other employment at a similar level. Following very rapid Africanization of the posts of Settlement Officer on the Kenya Million-Acre Scheme between 1964 and 1966, numerous prosecutions for embezzlement were brought against Settlement Officers.[3] Some African settlement staff, like some of their European predecessors, showed signs of being persons displaced from the normal and orthodox spheres of work and channels of promotion, and of being prepared to turn their hands to almost anything.

The impermanence of senior posts on many settlement schemes and the origins and motivations of those recruited to fill them combined

[1] N. S. Carey Jones, *The Anatomy of Uhuru: an Essay on Kenya's Independence*, Manchester University Press, 1966, p. 30.
[2] Gaitskell, op. cit., p. 325.
[3] See, for instance, the *East African Standard* (Nairobi), 20 July 1966.

to create a marked lack of staffing continuity. Gaitskell's observation that many field staff left Gezira, the turnover in field personnel at one time rising to 'a hundred per cent in twelve years'[1] serves to emphasize yet again the marked differences between Gezira and other schemes: for the continuity which appeared low to Gaitskell would have been inconceivably high for most other schemes. In most countries it was, in fact, only in the period before about 1950 that much continuity of staff was often achieved. Alvord, who started the Sabi Irrigation Projects, maintained his interest in them from 1928 until 1951.[2] The D.C. responsible for the Zande Scheme was D.C. from 1932 to 1946 and then Resettlement Officer from 1946 to 1950.[3] The development team which launched the Shendam resettlement operation remained unchanged from 1947 until 1951.[4] In all these cases, achievement was closely related to continuity. Gaitskell wrote of Gezira that 'personal continuity was of immense importance in the success of the project'.[5] Alvord's influence over the Sabi Irrigation Schemes was so great that they were largely his creation. It was only the high respect in which the D.C. of the Azande was held, as a result of his long association with the tribe, that made possible the achievement of a form of resettlement which was to the disadvantage both of the Azande and of the cotton producing and processing scheme it was intended to serve. In the case of Shendam an American observer attributed the 'healthy relationship' between the Colonial administrators and the Native Authority to staff continuity.[6] In contrast, from about 1950 onwards, continuity among the senior staff of settlement schemes was not common. The experience of Mwea was probably typical.[7] The Niger Agricultural Project was weakened after two years of operation by the departure, within three months of one another, of the Manager, Deputy Manager and Assistant Manager, together with the senior representative of the parent agency in Lagos.[8] One group farm in Uganda had five different Managers in two years. The insecurity sensed by expatriates and the high mobility of local staff together produced a rapidity of turnover which demoralized them and could reduce their activities to attempts to maintain a status quo. Only where, as on Mwea, the core processes of a scheme had already been reduced to routine and were well understood by both staff and settlers

[1] Gaitskell, op. cit., p. 104.
[2] Roder, op. cit., pp. 103–105.
[3] Pierre de Schlippe, *Shifting Cultivation in Africa, the Zande System of Agriculture*, Routledge and Kegan Paul, 1956, p. 20.
[4] Colonial Service Second Devonshire Course, Cambridge, 1952–53, op. cit., p. 13.
[5] Gaitskell, op. cit., p. 104.
[6] Schwab, op. cit., p. 55.
[7] See pp. 107-108 and Table 6.1 on p. 110.
[8] Baldwin, op. cit., pp. 90–91 and 194.

did this not have a serious effect. Even where young idealists and international experts were recruited the situation was not much better. Contracts for Peace Corps Volunteers and also for Uganda Group Farm Managers were normally for only two years, while British Volunteer contracts were initially for one year.[1] Besides, as Tickner has pointed out, the technical co-operation expert is not making a career, but interrupting one.[2] Not surprisingly, the motivation of such short-term senior officers has been directed more towards short-term technical achievement than towards the formation of lasting human relationships. Thus whereas before about 1950 continuity of staffing allowed the building up of paternal relationships, the later discontinuities combined with other factors to encourage greater emphasis on technical aspects of management.

Some characteristics of junior staff

The neglect of junior African staff in studies of development in Africa would be remarkable were it not for the difficulties of studying them. Although a certain amount of work has been done on African chiefs, little attention has been paid to the less visible junior staff working in the technical departments. Yet in many respects they occupy a crucial position. In field administration they often provide the continuity which is missing at the senior staff level. On settlement schemes they are often the main link in communication between senior staff and settlers. Much of the efficiency of the productive process of a settlement may depend upon their capacity to understand and interpret instructions, and to observe and report back on their effects. But junior staff have rarely been considered in any detail in studies of settlement. An attempt is therefore made here to describe the junior staff on Mwea.

The junior staff on Mwea were divided into two establishment categories: graded staff, and subordinate staff. The graded staff included Field Assistants, Clerks, Artisans and Drivers, while subordinate staff comprised Water Guards, labour headmen, watchmen and others. In addition the Settlement employed casual labourers on a daily basis. The graded staff were appointed with the approval of the Ministry of Agriculture and were required to pass tests in their particular skills, except that long good service by Field Assistants was taken as qualifying them for upgrading without a test. Some of the graded staff were transferable outside the Settlement, but few such transfers took place. The subordinate staff were direct employees

---

[1] The British Volunteer Programme has subsequently shifted to strong encouragement of somewhat longer periods.

[2] F. J. Tickner, *Technical Co-operation*, Hutchinson, 1965, p. 149.

145

of the Settlement and their employment was not subject to Ministry approval. Some of the graded staff were on pensionable terms, but none of the subordinate staff. Similarly graded staff were difficult to dismiss, whereas subordinate staff could be dismissed without difficulty. All, however, were subject to government regulations and terms of service.

Recruitment followed a changing pattern. Much of the early technical work was carried out by detainees with suitable experience, some of whom stayed on when drivers and artisans were recruited. Since a high proportion of detainees had worked in the European settled area, or in towns like Nairobi or Mombasa, there was a considerable pool of skills on which to draw. Of the Drivers, Field Assistants and Headquarters technical staff who were interviewed in mid-1966,[1] all but one of the nineteen who had begun work on the Scheme in 1955 and 1956 had been detainees, and brought to the Scheme experience as drivers (five), clerks (four), teachers (three), blacksmiths (three), and carpenters (two), while one had been a mechanic and one a mason.[2] They were all persons who had been displaced by the Emergency and who had travelled within Kenya. Without exception they were used to being employed rather than farming on their own account. Twelve had worked in Nairobi, ten in the European settled area, and four had been outside East Africa in the Army.[3] All but one said they originated outside Kirinyaga District. As time passed, however, the pattern changed, with a steady increase in recruitment of Gichugu and Ndia from nearby, this trend being most marked among subordinate staff. Thus of the eight Water Guards who started work on the Settlement in 1965 and the first half of 1966, five were Gichugu or Ndia. While the field of recruitment shrank geographically in this way it also moved towards younger men with better education. The changing fields of recruitment were interestingly preserved in the geographical distribution of field staff: Tebere, the first section to operate, had older and less well educated field staff, mainly Kikuyu from outside Kirinyaga District; Nguka was in an intermediate position; and the new Thiba section had the youngest and best educated staff, mainly from Kirinyaga District.

One of the most marked contrasts between junior and senior staff on Mwea lay in continuity. Just over half the junior staff interviewed in mid-1966 had served on the Settlement since before 1960, while

---

[1] The 100 junior staff interviewed break down into Drivers (23), Field Assistants (13), Clerks (5), Headquarters technical staff (14), Water Guards (20), and Supervisors and Labourers (25).

[2] These totals are not exclusive: for instance, the same person might be counted both as a clerk and a teacher if he had worked in both capacities.

[3] These totals are not exclusive.

in the same year senior staff were still in a state of rapid turnover.[1] Many explanations for this relative stability can be adduced. Job opportunities outside the Settlement were not good in the early 1960s. Most Field Assistants could not have expected to hold down equivalent jobs elsewhere; their qualifications were maturity and experience gained on the Settlement rather than formal training. Again, the seasonal work, with periods of intense activity and spells of relative idleness, may have been an attraction to many.[2] Habit and inertia may well also have been important. Status, prestige, and the clearly defined position in a hierarchy expressed in the tools of work or in transport—the Driver's tractor, the Field Assistant's motorcycle, the Water Guard's official bicycle—must all have counted. Perhaps most significantly junior staff, unlike senior staff, could perceive a career structure and opportunities for higher pay within the Settlement. Many had doubled, trebled, or quadrupled their salaries over the years, while the cost of living, although higher on the Settlement than off, had not substantially risen.[3] Another indication of the opportunities for advancement is indicated in Table 7.1.[4] From 1960 to 1966 the numbers of graded staff rose from 36 to 55 with the prospect of further increase as the Settlement expanded according to programme. The two most clearly perceived channels of promotion were for Water Guards, who aspired to become Field Assistants, and for labourers, who could save up for a private driving course and become Drivers. The prospects for the latter, if they gained their licences, were particularly good: of the fourteen Drivers interviewed who began work on the Settlement in jobs other than that of Driver, all but one had learned to drive since coming to the Settlement. At the same time the Settlement was increasingly tending towards self-sufficiency in the provision of staff, as the sons of tenants seized the opportunities for employment as they arose. Four of the five Water Guards taken on in 1966 were the sons of tenants and the fifth had a brother as a tenant. More new jobs went to relatives of tenants than to relatives of staff, but this is not surprising since at the end of 1966 tenants outnumbered junior staff by over 10 to 1.[5] But whether new employees were related to tenants or to junior staff, the tendency was the same: an integration and, it may be assumed, an

---

[1] In 1966 there were changes in the occupants of the posts of Manager, Accountant and A.A.O. in charge of maintenance (M.I.S. Annual Report 1966, p. 1.).

[2] The work of Drivers was strongly seasonal, but Field Assistants were busy throughout the year.

[3] The most spectacular rise of salary was from Shs. 55 per month to Shs. 935 per month, for the most highly skilled member of the headquarters junior staff.

[4] See p. 148.

[5] At the end of 1966 there were 1,588 tenants to 150 junior staff (M.I.S. Annual Report 1966, pp. 1, 2 and 5).

increased stability and continuity of the human factor in the Settlement.

TABLE 7.1

Mwea Irrigation Settlement:
Junior staff establishments

GRADED[1]

	1960	1966
Field Assistants	4	14
Clerks	5	6
Artisans	12	12[2]
Drivers	14	23
Other	1	—
	36	55

SUBORDINATE STAFF

	1960	1966
Water Guards	11 (1961)[3]	19
Others	(104)	76
	115[4]	95[4]

Sources: M.I.S. Annual Reports 1960, 1961 and 1966.

Notes:

[1] In 1960 this category of staff was described as 'A.C.S.' (African Civil Service), but the occupational groups were the same as with the later 'graded' staff.

[2] In 1966 this category was split into Artisans (8) and Tractor Supervisors (4).

[3] The 1960 Annual Report does not give the numbers of Water Guards.

[4] 115 is the 1960 total for subordinate staff. The decline from 115 to 95 is accounted for by the abolition of village sweepers, of whom there were 43 in 1960. There was, thus, an increase over the period in other categories of subordinate staff.

The permanence and stability of this integration must not, however, be exaggerated. Few members of the junior staff contemplated retiring on the Settlement. Their deep and powerful desire was to obtain their own land in order to gain security, food and money, and suitable work for their wives. The land cultivated by them on the Settlement consisted of small plots of unirrigated red soil, usually less than an acre, to which they held no permanent rights and which did not provide a basis for more than the unreliable cultivation of a few vegetables. In 1966, in response to an open-ended question about the future,[1] by far the most common preoccupation was with the acquisition and development of land, particularly among the better paid respondents—the Drivers, Field Assistants, headquarters technical staff, and Clerks. For them the Settlement had not become in any sense an encompassing way of life, from the cradle to the grave. Rather it was a valuable semi-permanent port of call which provided a living and which might make possible a later more secure existence on land elsewhere. But among some junior staff there was a desire to obtain a tenancy as opposed to continuing as an employee, a desire which as Table 7.2 shows, correlated inversely with salary. This is not a simple phenomenon, but it does suggest that the lower paid junior staff saw their opportunities more within the Settlement, through promotion or a tenancy, whereas the better paid junior staff, like the senior staff, tended more to see their long-term opportunities outside.

TABLE 7.2

*Mwea Irrigation Settlement: Junior staff
attitudes towards becoming a tenant or remaining
a Settlement employee (mid-1966)*

(Responses to the question: 'Would you prefer to become a
tenant or to continue as a Scheme employee?')

	Salary in shillings per month				
Preference expressed	Less than 100	100–199	200–299	300–399	400 Plus	Total responses
Employee	3	19	13	17	5	57
Tenant	6	14	8	5	—	33
No reply or Don't know	—	2	3	5	—	10
Total	9	35	24	27	5	100

Some of these features are to be found or expected in other settlement schemes. It appears general that junior staff are recruited from a wider field in the earlier stages of a scheme than in the later. The Niger Agricultural Project, for instance, recruited alien Ibo workers, but was put under political pressure to take on local Nupe.[2]

---

[1] The question was 'What do you hope to be doing in five years time?' Although unsatisfactory because of the imprecise position along the scale between a question about expectation and one about wishful thinking, the responses were of qualitative value.

[2] Baldwin, op. cit., p. 34.

149

Equally it may be expected that where a scheme continues as an organization over a period of time, there will be inter-marriage between staff and settlers, and jobs will increasingly be taken by those brought up on the scheme. Another stabilizing factor is the common tendency over time for junior staff to become settlers, and for settlers to perform roles that might otherwise have been undertaken by junior staff. On Upper Kitete, for example, the tractor drivers were themselves members of the settler community. Similarly on one Uganda Group Farm the Manager encouraged his staff to become farmers so that when government eventually withdrew the scheme would have a pool of skills on which to draw.[1] Much, of course, depends on the size and permanence of the organization. A small scheme cannot provide a career structure, except where it is part of a larger organization, and a short-lived scheme cannot provide security of employment except where staff become settlers. The general pattern is, however, clear. The extreme mobility observed among senior staff, subject as they have been to cross-postings, the disruptions of Africanization, and short-term contracts, has not been found among junior staff who are less likely to be posted and who have or develop local connections and interests.

Some characteristics of settlers

In discussions of settlement schemes the term 'selection' is often used to describe the process whereby settlers are found and brought to schemes. The word may seem to imply a choice on the part of a recruiting agency from among a number of applicants, but that is not necessarily the position. Some potential settlers constitute a problem a scheme is designed to solve: they are a group that has to be settled. Others may have to be compelled or enticed to join a scheme. Yet others may already be living on the land to be occupied by a scheme and are selected by incorporation. Finally, there are those who are indeed chosen from among a number of applicants. 'Selection' is used here in a broad sense to cover all these processes whereby potential settlers are identified and brought into contact with a settlement agency.

From the point of view of selection and the demographic characteristics of settlers, three types of scheme may conveniently be distinguished, though the boundaries between them in particular cases are not always precise: these types are 'mass', 'incorporative', and 'specific'.

Mass settlement schemes include moves of population resulting from acts of god or government, or combinations of these two types

[1] Simon Charsley, 'The Group Farm Scheme in Uganda: a Case Study from Bunyoro', op. cit., p. 19.

of displacement. Large-scale tsetse resettlements such as Anchau, and the resettlement operations resulting from the Aswan, Kariba and Volta dams are examples. Similarly, refugee settlements may include whole communities, though refugees are sometimes only one segment of the population from which they originate. With this qualification for refugees, the demographic characteristics of the people for whom mass settlement schemes are intended are typically those of an existing population.

Many schemes incorporate a population already occupying the scheme's land. These incorporative schemes include large-scale land reorganization involving changing settlement patterns, such as the Zande Scheme, where the whole Zande population was affected, and the Kenya Million-Acre Scheme in which the labourers of the former large farms were meant to be given priority in resettlement. Sometimes the population incorporated has unusual characteristics because of the history of the land. Thus at Rwamkoma in Tanzania the settlers incorporated from the people already on the land were exceptionally diverse ethnically and had few close relationships with one another because the area was a recently occupied frontier of settlement open to many groups.[1]

Specific settlement schemes differ from both mass and incorporative schemes in that there is deliberate recruitment or attraction of settlers with characteristics unrepresentative of a historical community. Frequently this specificness is a function of the expressed purpose of the scheme. The first settlers on Mwea were almost entirely 'landless whites' since the main purpose of the Scheme was to settle rehabilitated detainees. Nyakashaka in Uganda and the early Farm Settlements in Western Nigeria were designed for school leavers, and hence included a high proportion of young bachelors. The Sabon-Gida Settlement at Shendam in Northern Nigeria was for ex-servicemen. The high density plots on the Kenya Million-Acre Scheme were intended for people who were unemployed and landless. The Kilombero Schemes in Tanzania were for urban unemployed. In other cases, the characteristics of settlers were determined partly by special requirements. Farmers settled on low-density farms in the Kenya Million-Acre Scheme were required to have previous agricultural experience. For Upper Kitete in Tanzania a deliberate attempt was made to select men with useful skills, a need for land, and good character.[2] Sometimes the method of selection determined who was recruited. Where selection was compulsory, those selected were

---

[1] Rodger Yeager, 'Micropolitics, Persistence, and Transformation: Theoretical Notes and the Tanzania Experience' (mimeo), Paper to the African Studies Association of the United States, 1967, pp. 34–36.

[2] Personal communication from Garry Thomas.

M   151

persons who were for various reasons vulnerable to compulsion. For instance, some of the first twenty settlers at Kongwa in Tanganyika were recruited by means of pressure brought to bear upon them as tax-defaulters.[1] Nor was forcible recruitment limited to the colonial period. In 1967, 60 vagrants were reported to have been rounded up in Tanga in Tanzania, taken to a TANU Youth League Settlement, and threatened with corporal punishment if they left it.[2] In some other cases settlement schemes have been used as an opportunity for removing people who were considered undesirable: the first 'landless whites' at Mwea included some detainees whom Chiefs were not prepared to accept back into their home areas, and who could be compelled or forcefully persuaded to go to the Scheme; and in the Kiga Resettlements in Uganda some Chiefs saw the operation as a way of getting rid of stubborn men.[3] In these various instances, whether recruitment was voluntary or compulsory or some blend of the two, it is possible to trace the specific and unrepresentative characteristics of the settlers to the policies, methods and interests of the recruiting agencies.

With these specific settlement schemes, however, except where compulsion has been used, the characteristics of the settlers have depended not only on the agency but also on who was available for selection. The less popular the scheme, the lower the calibre of settler who would come forward. An extreme case was the Niger Agricultural Project where a failing scheme received twenty-seven new settlers in its third year, of whom eight were 'crocks', including four lepers.[4] A similar but less extreme tendency was noticed on Mwea in 1965 as a result of the recruiting environment becoming 'fished out'. From the Management's point of view the ideal settler was a family man aged between 35 and 50.[5] Of the 148 new tenants selected for the Scheme in 1965, however, 86 were under 35, 34 were over 50, and only 28 were in the ideal age group.[6] In both these instances selection was constrained by limited supply.

Where the catchment for settlers has been broader, and compulsion has not been used, there has been a striking tendency to attract persons who have been mobile and who have experience of wage employment. Among the Sabi Irrigation Project farmers interviewed by Roder only a few lacked any experience of working in an urban area.[7] Rigby has observed that many of the tenants at Kongwa had

[1] Rigby, op. cit., p. 16.
[2] *Reporter* (Nairobi), 24 March 1967.
[3] Katarikawe, op. cit., p. 4.
[4] Baldwin, op. cit., p. 61.
[5] M.I.S.C. minute 26/65 of 13 July 1965.
[6] Memorandum by Acting Manager, c. May 1965.
[7] Roder, op. cit., pp. 92–93.

worked for wages in towns and plantations.[1] At both Kabuku and Amani in Tanzania, most of the candidates for settlement were ex-estate workers.[2] At Rwamkoma, of the first settlers, who were recruited from a spontaneous frontier of settlement, 27% had been employed in the past by the Tanzania and Kenya Governments, and 77% had worked in private economic organizations or participated in large agricultural co-operative societies.[3] The evidence from Mwea is also significant. Technically only landless men were eligible for settlement, which biased selection towards those who had had to look for work in the wage sector. Indeed, in a 1966 survey[4] of ninety-nine tenants, 57 of whom were Kikuyu and 42 of whom were from Kirinyaga District, only one respondent positively recorded that he had never been employed.[5] Although the Kikuyu tenants presented a denser assembly of special skills than the Kirinyaga tenants,[6] the life histories of the latter suggested that even when recruitment was from a relatively confined area settlers possessed experience unrepresentative of the population as a whole. Mwea, like many other specific settlement schemes, was a home for migrant wage-earners.

Despite these similarities among settlers, a common characteristic of settler communities has been their heterogeneity. Where this has been ethnic, problems of a social and political nature have been associated with tribal and cultural differences. The divisions between Kikuyu *ahoi* immigrants and local Kirinyaga on Mwea are typical of schemes with ethnically mixed populations. The Perkerra Irrigation Scheme in Kenya has similarly been divided between Tugen immigrants and local Njemps, considerably complicating problems of management and tenant representation. In Sabon-Gida in Northern Nigeria the prolonged administrative problems of the fifty settlers have been attributed to their consisting of three ethnic groups (with 30, 10 and 10 settlers respectively),[7] a situation contrasted with the

[1] Rigby, op. cit., p. 47.

[2] Personal communication from John Nellis (1966).

[3] Yeager, op. cit., p. 36.

[4] The UNICEF/Makerere Farm Innovation Survey carried out during the summer of 1966, by students in the Department of Agriculture, Makerere University College, under the direction of Jon Moris. 100 tenants were interviewed but one interview has been rejected as unsatisfactory.

[5] There were in addition three inconclusive responses.

[6] Comparative figures for selected occupations were:

	Kikuyu (57)	Kirinyaga (42)
House servant	12	5
Mason	9	1
Driver	5	3
Carpenter	3	2

[7] Schwab, op. cit., p. 47. The Yergam minority refused to pay tax to the headman, who was one of the Ankwe majority group.

nearby Mabudi settlement which was entirely homogeneous and in which village conflicts did not arise.[1] In the Volta River Resettlements the policy of grouping evacuees into towns made it inevitable that different ethnic groups were brought together, leading to a complete split of settler institutions in one extreme case,[2] and elsewhere creating communities with physical but not cultural or social unity.[3] As at one time in Israel, so also in Tanzania it was a matter of political policy that settlers should be of mixed ethnic origin;[4] but in Israel this policy was abandoned in face of the practical social difficulties encountered,[5] and in Tanzania too, selection in practice sometimes sought to limit settlers to one cultural group.[6,7]

Managers' and settlers' initial attitudes

Managers and settlers have brought to settlement schemes not only very different ethnic and occupational characteristics, but also different attitudes and views of the schemes. These attitudes and views have varied according to many factors, not least personal idiosyncracies. Nevertheless, certain general tendencies can be observed. It is impossible to divide these attitudes and views neatly into those which have been brought to the schemes, and those which have developed through the scheme experience. But the discussion here is of those which have been more closely associated with the presenting cultures and interests of managers and settlers, while reserving for the next

[1] Ibid., p. 67.

[2] Tonkor-Kaira, which in August 1965 had two separate Town Committees, one for New Tonkor, and one for New Kaira.

[3] Interviews with Town Managers and visits to settlement towns (August 1965). At Danyigba, with a population of just over a thousand, there were said to be no less than 8 active gong-gong beaters (town-criers) for 8 of the 10 groups.

[4] Nikos Georgulas, 'Structure and Communication: a Study of the Tanganyika Settlement Agency', unpublished D.S.Sc. dissertation, Syracuse University, 1967, pp. 128–129

[5] Alex Weingrod, 'Administered Communities: Some Characteristics of New Immigrant Villages in Israel', *Economic Development and Cultural Change*, Vol. 11, No. 1, October 1962, p. 73.

[6] For instance at Upper Kitete all 100 farmers were Iraqw, except 7, 4 of whom had Iraqw wives, and all of whom had been long-term residents of the area (Garry Thomas, 'Effects of New Communities on Rural Areas—the Upper Kitete Example' (mimeo), paper presented at the Seminar on Rural Development, Syracuse University and University College, Dar es Salaam, 4 to 7 April, 1966, p. 4).

[7] The question of how much homogeneity is desirable is not simple. However, as with some democratic party systems, political stability and effectiveness in settlement communities may tend to vary inversely with the number of organized factions. For an interesting discussion of this point based on a comparison of seven Moroccan settlements in Israel see Alex Weingrod, *Reluctant Pioneers, Village Development in Israel*, Cornell University Press, Ithaca, 1966, pp. 179–188.

chapter discussion of those attitudes and views which seem to have emerged more as a result of interaction and adaptation in the scheme situation.

The most conspicuous attitudes of many expatriate managers were related to their European cultural background, which predisposed them to set a high value on hard work, predictability, achievement and exact measurement. They were preoccupied with imposing order, as they saw it, upon a state of affairs which if not chaotic was certainly unorganized. They might spend much time drawing maps and graphs and making calculations. They often displayed a strong desire for physical arrangements—of housing, fields, and roads—in straight lines. The D.C. who resettled the Azande considered that 'the mathematical precision' of the resettlement was 'the scaffolding on which we hope to build a settled population of peasant farmers with sufficient share of the best land available, and living in an accessible and rational manner, not as beasts in the wilderness'.[1]

Of Uganda settlement schemes, Belshaw has observed that planned resettlement might be intrinsically desirable to the official mind, and that there has often been an excessive preoccupation with orderliness to the detriment of, or irrelevant to, the welfare of settlers or the efficiency of resource use.[2] Traditional shifting agriculture was seen by expatriates as unstable perhaps not only because of the damage it did when intensified, but also because it was untidy and consequently difficult to inspect or control. It may well be that the recurrent drives in colonial agricultural policies for the transformation of shifting agriculture, the planning of holdings, the demarcation of boundaries and the settlement of population in some geometrical and predictable pattern owed at least as much to the psychological needs of the expatriates who promoted them as to the requirements of the situation. Visible agricultural order may have reassured expatriate agriculturalists and administrators, allowing them to feel that they had tamed and converted the primitive chaos of the dark continent, in much the same way as an early missionary might have felt safety and pride in the sight of his African parishioners in Victorian Sunday clothes. For the satisfaction of this need settlement schemes, with their captive populations, provided an ideal opportunity.

One aspect of settlement schemes that might seem related to this preoccupation with orderliness has been the almost universal allocation of equal resources to settlers, at least at the outset at the time of settlement. The standardization of four-acre holdings at Mwea, 40-feddan tenancies of Gezira and 15-feddan tenancies of Managil,

---

[1] Letter, Zande District file No. 1.C.8, 15 October 1949, quoted by Reining, op. cit., p. 132.
[2] Belshaw, op. cit., p. 2, footnote.

6-acre holdings of Nyakashaka, and the recommendation for 10-acre plots for Mubuku,[1] are examples of a uniformity widely found within any one settlement scheme. Indeed, in the Settlements of the Kenya Million-Acre programme, this principle was carried to the sophisticated length of planning different acreages for different plots in an attempt to achieve equal income-earning and therefore loan-repaying capacity.[2] It might be supposed that there was an ideological reason for this emphasis on equality; but many of those responsible for planning schemes were, personally, conservatives. Only with post-independence settlement schemes was socialist egalitarianism sometimes a strong influence, as with the Tanzanian schemes in which it was intended that all settlers would have equal shares of good and poor soils.[3] More often administrative convenience was the reason for standardization. Giving all settlers equal resources reduced grounds for complaints of favouritism and simplified the allocation of plots. Further, there could be book-keeping advantages in this uniformity. Yet another factor was the ignorance on the part of the Manager of the attributes of the people who were to be settled. The less known the people were, the stronger was the argument for standardization, and vice versa. In the Volta Resettlement operation, where chiefs were conspicuous and known to resettlement staff before the move, their status was recognized in the allocation of special houses or more than one house. More generally, managers regarded settlers as uniform phenomena, to be understood and manipulated according to some preconceived idea of settler human nature, as Africans who could be discussed in terms of 'the African', as shadowy figures who could be expected to act identically. Such attitudes were not necessarily racial. They can be found in any institution which receives intakes of people into subordinate positions.[4] Convenience of administration and ignorance of individual differences combined to make the desirability of standardized treatment so obvious as not normally to require justification, especially as the order and uniformity that resulted were reassuring and satisfying to the managers.

Managers' images of schemes, ideas of their own roles, and attitudes towards settlers were influenced by their own origins and by the dominant political patterns of the time. During the colonial period the wandering men who often became managers were usually

[1] Agrawal, op. cit., p. 5.
[2] C. P. R. Nottidge and J. R. Goldsack, *The Million-Acre Settlement Scheme 1962–1966*, Department of Settlement, Nairobi, 1966, pp. 15–21.
[3] *Rural Settlement Planning*, issued by the Rural Settlement Commission, Vice President's Office, Dar es Salaam, 1964, p. 3.
[4] See Erning Goffman, 'The Characteristics of Total Institutions, in Amitai Etzioni, ed. *Complex Organizations, a Sociological Reader*, Holt, Rinehart & Winston, New York, 1962, pp. 312–340.

in sympathy with a colonial-paternal set of attitudes. Later, near and after independence, especially with the young idealists, a democratic-advisory set of ideas came to bear on settlement schemes, but was moderated by what may be described as technocratic-disciplinary approaches to management, especially as pursued by foreign aid experts. Although it could be misleading to stress the association of these three sets of attitudes with the three categories of manager, there was nevertheless some correlation.

In the colonial era, managers' images of schemes and of their own roles were fashioned in a colonial mould. The Governor ruled 'his' colony. The D.C. ruled 'his' district. The Manager of a settlement scheme might naturally seek to make 'his' scheme a colony or a district in miniature. In this the territoriality of colonial officials and of colonial managers came into play. The battle between the Administration and the Agricolas for Mwea can be interpreted as a war between rival colonists for the right of ownership of the Scheme. The battle brought into the open and made explicit attitudes which were probably latent in many settlement managers. On Mwea the second Manager, in seeking to be in a position to control 'all happenings' on the Scheme, to acquire magisterial powers, and to exercise a high degree of autonomy, was expressing aspirations which may well have been very general. Indeed, in so far as a manager's behaviour was possessive, authoritarian, or paternal, it was only following a long recognized human tendency. Clear territorial boundaries, physical isolation and a community of 'subjects' gave a manager a position in which he might easily come to regard himself as a sort of chief with sovereignty over his settlers, the lord of an island in which the settlers would play Caliban to his Prospero or Man Friday to his Crusoe. These attitudes were reflected in managers' views of settlers. The Resettlement Officer for the Zande Scheme regarded the Azande as good-natured and amenable, but he believed that they would only produce the cotton required if they were adequately controlled and supervised.[1] 'The personal touch and the paternal manner', he wrote, were desirable in dealings with them.[2] The Europeans in charge of the Scheme also believed that the Azande needed protection from the evils of urbanization and of excessive quantities of money.[3] On Gezira there was a similar, but less patronizing set of attitudes which stressed the importance of patience in dealing with tenants and of taking an individual interest in them. These paternal attitudes sometimes reflected managers' views of themselves as wise and benevolent and knowing what was best for the settlers, and of settlers as simple,

---

[1] Reining, op. cit., p. 133.
[2] Ibid., p. 132.
[3] Ibid., p. 190.

innocent and weak-willed. But Gaitskell's comment carries conviction when he writes that a critic of paternalism would be wrong to sneer at the early Gezira approach since as long as a father-figure was important, confidence was derived from this sort of behaviour.[1]

The democratic-advisory set of ideas and attitudes evolved with moves towards and beyond political independence, and as colonial wanderers gave way to young idealists and Africans as managers. Partly it reflected a seepage into schemes of political ideas. Schemes could be regarded as nations in miniature or as democratic local governments in miniature, embodying a degree of self-government through settlers' co-operative organizations, to which managers should be merely advisers. Since at a national level government had been handed over from an autocratic European colonial power to a democratic African independent nation,[2] so on settlement schemes it was felt that the expatriate manager should prepare to hand over control to an elected settler body. It may be significant that in Tanzania expatriate managers were less inclined than their African successors[3] to believe that this devolution could be carried out quickly. For expatriate managers were experiencing on a small local level the stresses of abandoning power which had earlier affected individuals within the colonial government, and did not find it easy to give up their positions. Nor did they always find it easy to accept the political view of settlers as responsible individuals capable of operating a co-operative system successfully. However, this varied considerably by type of scheme. With schemes where complex central services were provided, managers tended to consider devolution difficult or impossible, but a less pessimistic view was taken where managerial units were small, and co-operation was limited to marketing.

With the colonial model disbanded, the paternal image tarnished, and the co-operative democratic orthodoxy widely accepted, managers were driven to defending their positions by invoking more strongly the need for controls on technical grounds. Coinciding with this shift the foreign aid experts who sometimes came in to manage schemes often appreciated local conditions less than their predecessors, were less interested in personal relationships, and regarded themselves as technicians concerned, rather like managers of commercial estates or plantations, only with securing high production and high returns. To them settler representative bodies were analogous to trade unions, and settlers to labourers. They demanded high disciplinary powers to ensure the presence of settlers on the job, and easily applicable sanctions (including eviction) for failure to comply

[1] Gaitskell, op. cit., p. 229.
[2] Except in the case of Rhodesia.
[3] Personal communication from Robert Myers and John Nellis.

with instructions. At the same time the more complex schemes that were being introduced provided genuine justification for discipline on technical grounds. The high degree of authority previously vested in the white father figure could now be claimed on the basis of technological imperatives.

These changes in managers' attitudes were paralleled in those of settlers. Any discussion of this subject is complicated by the number of variables involved, and by the absence of any sociological framework of settlement theory upon which to hang the presentation. It is, however, possible to see that settler attitudes to settlement situations can be separated into two historical periods: an earlier dependent period, corresponding with the colonial-paternal approach on the part of management; and a later independent period corresponding with democratic-advisory and technocratic-disciplinary approaches.

In the colonial period, after colonial rule had been accepted but before tribal values had been seriously weakened, government was regarded as an incontrovertible fact, as part of the ecology,[1] as an essential part of the universe.[2] A resettlement operation was accepted as an act of god, as a natural event like a drought or flood which it was useless to try and resist. This would seem the most plausible explanation of the easy acceptance of compulsory resettlement by the inhabitants of the Gezira plain which has puzzled Gaitskell,[3] and of the amenability of the Azande to resettlement and compulsory cotton cultivation.[4] At this stage relationships between settlers and managers were characterized by the reciprocal dependences of paternalism, of master and servant, or of patron and client, relationships which, as Mannoni has shown, were important both to the European colonial and to the dependent African.[5]

Later, however, as tribal systems were transformed through the impact of the modern world, settlers' attitudes became more independent. Settlement no longer appeared as an act of god but as an act of man. Government was no longer seen as invincible and inevitable. A settlement policy might be altered as a result of political pressure, and settlement itself could be regarded as but one of a number of alternatives between which a person could choose. In these historical circumstances, settlers claimed a greater say in policies, and were more concerned with maximizing their rewards from the schemes in which they participated.

[1] Paul Spencer, *The Samburu: a Study of Gerontocracy in a Nomadic Tribe*, Routledge and Kegan Paul, 1965, p. xx.

[2] Gustav Jahoda, *White Man, a Study of the Attitudes of Africans to Europeans in Ghana before Independence*, Oxford University Press, 1961, p. 111.

[3] Gaitskell, op. cit., pp. 89–90.

[4] Reining, op. cit., p. 134.

[5] Mannoni, op. cit., p. 81 and *passim*.

In both periods settlers' initial attitudes to management and to schemes were strongly conditioned by the manner of their selection. Where selection was unavoidable, either because of compulsion or because of other circumstances, problems often arose from resentment. The fact that only two of the first twenty settlers at Kongwa remained for a second year suggests dissatisfaction connected with the compulsory manner in which some of them were recruited.[1] Similarly, refugees sometimes objected to attempts to resettle them compulsorily.[2] Where a population was displaced by an act of government, such as building a new town or creating a lake behind a dam, the position was less clear. It was usual for compensation to be offered as an alternative to official resettlement, but the terms of compensation could be so unfavourable that, in accordance with the wishes of government, few people preferred them to being resettled officially. In the Tema resettlement operation in Ghana, government actively discouraged the people from accepting compensation, and in the end only two householders out of a population of 12,000 people preferred it to resettlement.[3] Similarly, in the Volta operation, only 13% of evacuee families opted for compensation.[4] Although the possibility of making such an option must have reduced resentment, many of those who were resettled could none the less feel that they had been caught up in a process beyond their control and become apathetic and fatalistic.

Where a government was not responsible for displacement and did not compel resettlement, a settler's or potential settler's attitudes could be interpreted in terms of a balance of forces between his home environment and the scheme. These forces can be described as push, pull, retaining and repulsing. Commonly, push forces have been population pressure, landlessness, unemployment, and social conflict, often combined in various ways. Pull forces have included economic opportunity, ownership of land, security, modernity and welfare. Retaining forces have often been economic interests and social relationships in the home environment. Repulsing forces have included unfavourable perceptions of a scheme, and fear of control, indebtedness, and disease. The desire or reluctance to become a settler has been the resultant of these four forces, and they have continued to operate after settlement in determining whether a settler remains or leaves, and in influencing his behaviour on a scheme.

[1] Rigby, op. cit., p. 16.
[2] Rachel Yeld, 'Implications of Experience with Refugee Settlement', *East African Institute of Social Research Conference Papers*, 1965, p. 12.
[3] Amarteifio and others, op. cit., pp. 7–8.
[4] G. W. Amarteifio, 'Social Welfare', *Volta Resettlement Symposium Papers*, op. cit., p. 85.

Stability of settlement has been achieved most noticeably in schemes which have combined strong natural, that is non-government, push forces with a lack of repulsing forces. The Sukuma, Shendam, Kigezi and Chesa Schemes provide examples where a strong push in the form of population pressure has been combined with limited controls for soil conservation, very little subsidy or assistance, and only minor changes, of a largely or entirely voluntary nature, in agricultural practices. In all these schemes settlers had to build their own houses and to carry out all or part of the clearing of the land. Woods's conclusion that the Native Purchase Area Schemes in Rhodesia succeeded because they were neglected can be applied to all known schemes of this type.[1] In contrast, from the point of view of stability of settlement, the most unsuccessful schemes have been in areas where push forces were weak and repulsing forces strong. The Niger Agricultural Project, which was not a response to any local need for settlement and which acquired a reputation for tough discipline, is a classic example. Repulsion appears to vary particularly with the degree to which there is 'scheme'—to which, that is, life on the scheme is perceived as controlled and directed.

Settlers' views of schemes and purposes in joining them have differed considerably. On loosely controlled schemes where there has been free recruitment, management and settlers have both tended to see the scheme as a means of providing security, subsistence and income to the settlers. With more ambitious and more closely controlled organizations there has almost always been a divergence at first between management and settlers in their ideas of the purposes of the scheme. Often such schemes have been regarded by settlers with suspicion. Some have seen them as government institutions where the inmates are not free, where, as on Mwea, a man might 'serve the whites forever'. Others have regarded them as government or European ruses for acquiring land.[2] The Azande felt that resettlement and the cultivation of cotton were compulsory work for the government.[3] Often, too, schemes have been regarded by settlers not as places for settlement as intended in the official rationale, but as institutions where a limited stay could be used for a specific end. The Kiwere Pilot Village Settlement was used by some settlers as a training school, where they went 'to learn tobacco'.[4] More common has been a tendency, associated with the occupational experience of the settlers, to treat schemes as estates or industrial firms on which they

---

[1] Woods, op. cit., p. 16. Settlement on the Chesa Scheme began in 1957, and by August 1964 only approximately 4% of the settlers had left (ibid., pp. 8–9).

[2] Roder, op. cit., p. 78. Colson, op. cit., p. 2. This was also true of Mwea.

[3] Reining, op. cit., p. 113.

[4] Personal communication from Robert Myers.

can work as migrant labourers for target sums of money and then leave: for instance, Baldwin concludes that many of the young settlers on the Niger Agricultural Project were using the scheme as an alternative to seeking work elsewhere;[1] Lord has suggested that the high rate of turnover at Nachingwea, roughly 50% of all tenants in their first two years, resulted from target-working;[2] and Rigby has found at Kongwa that settlers valued the scheme not so much for the cash income it provided as for the opportunity to accumulate cattle with which, once they had enough, they hoped to leave.[3] In Eastern Nigeria some settlers saw schemes so much as extensions of government that they regarded themselves not as farmers but as civil servants, and demanded monthly salaries and permanent and pensionable terms.[4] In all these cases settlers' initial attitudes were at variance with official intentions. Even where both manager and settler regarded the scheme as not unlike an estate, the manager was obliged to consider permanence of settlement as one of his major objectives while the settler might simply regard himself as a temporary employee.

The evolution of individual schemes can be understood partly as a dialogue between these disparate elements, these two teams, as they tended to become: the staff, especially the expatriate managerial staff; and the settlers, who were always African. Their different cultures, purposes, permanence and roles predisposed them to misunderstand one another and to disagree. One effect of the social distance and cultural differences between them was to tend to limit contact to matters in which they had a clear common interest. The managers were partly judged by the success of 'settlement', and therefore were concerned with settler satisfaction, or at least settler retention by the scheme. Settlers had an interest in high returns and except when they were target-working might be satisfied and retained by maximizing those returns. This in turn coincided with managers' professional interest in production. Indeed the historical trends towards schemes with more complex economic systems, managers who took a more technocratic and less paternal view of their work, and settlers who were more independent and cash-conscious in their attitudes, converged upon the economic process as the central concern in a scheme. Differing though they did in many other respects, staff and settlers could find a common interest in the economic system.

[1] Baldwin, op. cit., p. 161.
[2] Lord, op. cit., pp. 96–97.
[3] Rigby, op. cit., pp. 51–52.
[4] Raymond Apthorpe, 'Land Settlement and Rural Development: Part I, (pp. 1–4): Some Definitional and Conceptual Problems, Models and Phases' (mimeo), paper to University College, Nairobi, Conference on Education, Employment and Rural Development, 1966, p. 2.

CHAPTER 8

Staff-Settler Organization

Staff-settler relationships are, of course, influenced not only by the different characteristics and attitudes with which the two groups come to settlement schemes; they also evolve in response to scheme situations. In these the initiative, at least at first, comes more from staff than settlers. In order to pursue their aims of achieving settlement and production, managements devise organizations, methods of communication, and systems of rewards and sanctions, in order to reach down to, activate and control settlers. For their part settlers adapt to their situations and seek to influence management in what they perceive to be their interests. The result is an interplay between staff and settlers in which adjustments are made by both sides and as a result of which the organization of a scheme changes over time. These various aspects of scheme organization and functioning, as they are found in continuing schemes, will be considered in turn.

Managerial imperatives

Staff organization on settlement schemes varies considerably. Three principal variables appear to affect it, and these are strongly related to one another: size of managerial unit; length of scheme life-cycle; and technical requirements of the economic process. Where the managerial unit is small, where a scheme has a short life cycle from initiation to staff withdrawal, and where the economic process is not technically difficult (situations which tend to occur together) staff organization is typically simple, unspecialized, and short-lived. Where, however, the managerial and decision-making units are large, where schemes have long or indefinite lives, and where the economic system is technically difficult (again situations which tend to occur together) staff organization is typically complex, specialized, and continuing. Even if there is devolution of authority from staff to settlers, this is likely to preserve the complexities of the original.

Despite these polar distinctions there is a widespread tendency for settlement schemes to break down into manager-sized units, that is, into units of a size and complexity which can conveniently be handled

163

by an agricultural officer. These may be geographically distinct 'island' schemes, such as Ilora, Nyakashaka, or Upper Kitete, or parts of a larger scheme, such as the blocks of Gezira, the individual Settlements of the Kenya Million-Acre Scheme, and the sections of Mwea. The Manager of Mwea can be regarded in a sense as an overmanager with the field A.A.O.s in charge of sections as managers under him. Similarly on the Kenya Million-Acre Scheme, Senior Settlement Officers (overmanagers) have been in charge of a number of Settlement Officers (managers) each with his own individual Settlement. On the Million-Acre Scheme and Mwea specialized services have been provided at the overmanager level, but even so the junior staff below the manager level have been functionally differentiated: on the Million-Acre Settlements there were both Agricultural Assistants and Veterinary Scouts under each Settlement Officer; and on Mwea as we have seen there are Field Assistants responsible for tenants' agricultural performance, and separate water control staff responsible for the operation of the irrigation system.

The form and functioning of staff organization in the early stages of a scheme may be geared to construction, settlement and welfare. Later, however, they are explicable largely in terms of perceived economic imperatives.[1] What the imperatives are depends on the scheme. On the Kenya Million-Acre Scheme, where loan repayments have been a central concern of management, financial imperatives have been prominent, and accounting staff important. On Nyakashaka, where there is a strict imperative that plucked tea be delivered by the settlers to a central point at a certain time in order that it reach the processing factory within six hours of plucking, a focus of organizational attention has been the scheduling and punctuality of tea deliveries. On Nachingwea, where mechanized services were provided, the supporting and operating organization of maintenance staff and drivers was conspicuous,[2] as also on Mwea. Indeed, mechanization typically generates many imperatives, most notably the scheduling of other operations by settlers and attendance at certain times in order to carry out activities subsidiary to mechanized operations. An industrial situation is created in which a degree of regimentation is required by the process. The effect on management controls and style is marked. The manager of a scheme where farm decisions and activities are decentralized can afford to advise and educate rather than control and compel; the manager of a mechanized scheme, however, is driven by the inexorable imperatives of tractors, timing, and economic performance into seeking to command and instil discipline.

[1] See p. 114 for a discussion of some perceived economic imperatives on Mwea.
[2] Lord, op. cit., pp. 140–141.

It would be naive to suppose, however, that managerial organization and controls derive only from economic imperatives. Power is attractive for its own sake. Furthermore, some schemes have sets of rules and styles of management which have a totalitarian appearance. Especially with controlled irrigation schemes, like Mwea, the exacting requirements of the formal rules and the close attention to discipline suggest that the origins of the organization and its functioning owe something to psychological drives for power on the part of staff as well as to economic imperatives.

In this connection an intriguing question arises from analyses by Wittfogel[1] and Gray.[2] Wittfogel has sought to derive oriental despotism from the requirements of large-scale irrigation. Gray has suggested that the powerful hereditary oligarchy which allocates and controls water use among the Sonjo, an irrigation-based society in Tanganyika, has evolved as an ecological adaptation to the requirements of 'minute planning and continual supervision' in operating an irrigation system.[3] The question is whether there is an imperative or a number of imperatives associated with irrigation which require an authoritarian organization; whether in fact the management of an irrigation scheme is similar in origin and style, *mutatis mutandis*, to the totalitarianism of Wittfogel's hydraulic regimes or the powerful elders described by Gray.

At first sight Mwea fits these patterns. A high degree of control is exercised by the Management over the tenants. Their presence is required at certain times and their correct performance of a cycle of operations is subject to close surveillance. The Irrigation Rules can be used to require or prevent a wide range of activities, and include provision that a tenant should 'devote his full personal time and attention to the cultivation and improvement of his holding'.[4] A carefully graded series of penalties is applied to tenants who fall short of the standards required. Furthermore, water distribution is centrally controlled, and the temporary denial of water to tenants has on occasion been used as a collective inducement to improve performance. Eviction, equivalent to Gray's denial of water to an individual in that it makes crop-growing impossible, has been used as an ultimate sanction. Mwea can, thus, appear as a species of a genus which includes hydraulic despotisms and irrigation oligarchies.

[1] Karl A. Wittfogel, *Oriental Despotism: a Comparative Study of Total Power*, Yale University Press, New Haven, 1957.

[2] Robert F. Gray, *The Sonjo of Tanganyika: an Anthropological Study of an Irrigation-based Society*, Oxford University Press, 1963, especially pp. 55–61 and 166–171.

[3] Ibid., p. 170.

[4] *The Trust Land Ordinance*, cap. 100, *The Trust Land* (*Irrigation Areas*) *Rules, 1962*, section 8(1a).

These resemblances are, however, superficial. In both origin and continuation the controls on Mwea differ from those of the Wittfogel and Gray models. Wittfogel's thesis rests strongly upon the need for highly organized labour for the construction and maintenance of major irrigation systems, which need, he argues, has historically been associated with despotism.[1] But on Mwea construction was carried out largely by the M.O.W. and contractors,[2] and although minor feeders and drains are maintained by tenant labour, the large works (the major channels and drains) are the responsibility of the Management and do not require a large labour force. Gray's model is equally different. The power of Gray's elders derives from a combination of factors among which seasonal shortages of water are important: the scarcity of water and its consequent value give power to those who can allocate it or cut it off.[3] On Mwea there has never been any shortage of water, and were the sanction of denial of irrigation water to tenants to be used widely it would defeat the Management's objectives, since success is judged largely by production, and crops cannot be grown without water. The close control over tenants' activities at Mwea is genetically completely different from either that of a hydraulic regime or that of a hereditary irrigation oligarchy, for it originates neither from needs of construction and maintenance, nor from needs of allocation and distribution, but from imperatives of production.

The irrigation system does, however, impose some constraints. Tenants are required to weed and clean minor feeders and drains, some communally and others individually. The visibility of failures to carry out such clearing makes close control easier.[4] Further, these clearing activities, coming as they do immediately after harvest and payout, and not appearing closely related to production and returns, conflict more strongly with tenants' inclinations than does cultivation work. Nevertheless, on Mwea the irrigation system itself does not make many demands upon the tenants. Water management, in theory at least, is the sole responsibility of the water control staff under each A.A.O. in charge of a section and neither tenants nor Field Assistants are meant to alter water levels, nor in any way interfere with water distribution. The system has been designed partly to protect the tenant from any possibility of a Field Assistant and a Water Guard combining to deny him proper service in the form of water. The

---

[1] Wittfogel, op. cit., *passim*, especially pp. 22–29.
[2] Mass detainee labour was used in the early stages, but this does not affect the argument since that was a reflection of a passing phase in the total society, not, as in a hydraulic state, a permanent feature.
[3] Gray, op. cit., pp. 58–61.
[4] See Plate 5.

intention has been that if a Water Guard fails to provide water as required, a tenant has an alternative line of complaint through his Field Assistant, whose interest is in high production and therefore in effective water distribution to the tenants under his charge. Although in practice tenants have moved water, and Water Guards have worked to Field Assistants to some degree, this balance of power, and also the possibility of a direct appeal by a tenant to an A.A.O., have mitigated the despotic potential of the situation.

It has, indeed, been mechanization more than irrigation which has generated imperatives requiring close control and regimentation. The tractor teams rotavate successive areas according to a schedule. Before they come tenants are required to have cleared the ratoon[1] crop from their fields, and when rotavation takes place they are responsible for ensuring that the metal bridges used by the tractors to cross the bunds are carried from field to field.[2] More important it has been mechanization which has made possible the synchronization of cultivation activities, and consequently their closer control. In the words of one of the Mwea reports:

> The introduction of mechanical cultivation and the consequent need to adhere to a rigid programme allows very little latitude for the vagaries of individual tenants who do not fall in step with the majority. A lack of firmness in dealing with refractory individuals causes a breakdown in the programme and damage to the vast majority of tenants who are pulling their weight.[3]

On other mechanized schemes, too, the technical and economic imperatives of tractor use determine the nature and degree of the controls exercised by staff over settlers. For a Uganda Group Farm, Charsley has pointed out that mechanization means that the pattern of the farmer's activity is rigidly fixed:

> The crops he will grow, how much of each, when they will be planted, how they will be looked after, how and when he will dispose of them, are all decided without reference either to the resources he has available or to his own ideas or inclination . . . the machines demand that everyone conform to a fixed pattern.[4]

Indeed, wherever mechanization is only partial and is complementary to a settler's manual labour, it reduces his freedom of decision and action.

---

[1] The rice shoots which grow up again after the first crop has been cut.
[2] See Plate 3.
[3] M.I.S. Quarterly Report for July–September 1962.
[4] Simon Charsley, 'The Group Farm Scheme in Uganda: a Case Study from Bunyoro', op. cit., p. 23.

Other imperatives affecting the degree of control exercised by a management derive from the requirements of marketing. On Mwea marketing has had less effect than mechanization on the style and detail of the Settlement's operation. Tenants who deliver rice that has been inadequately winnowed have to clean it before it is accepted, but this is not felt as a major constraint. Similarly, the delivery of overdry paddy was a serious problem only until crop advances provided adequate incentives to ensure early deliveries. On Mwea it has been the cycle of production that has determined the timing and pattern of marketing more than the requirements of the market that have determined the manner of production.

As in the examples given above, managerial controls on settlement schemes can usually be explained in terms of requirements of the economic process, whether these are associated with finance, cultivation, mechanization, irrigation, or marketing. Power for its own sake may well be a motive among managers and junior staff in seeking to impose controls, but such controls can generally be justified in terms of perceived economic imperatives and thus of the interests of the settlers. That the settlers have not always seen matters in this light results from several factors, not least the difficulties of staff-settler communications.

Staff-settler communications

Many of the difficulties of staff-settler communications can be attributed to cultural differences. The most obvious of these is language. Among expatriates subject to rapid postings, there was little incentive to learn a local dialect,[1] and there was always a danger of learning just enough to multiply misunderstandings. Some communicated in a crude *lingua franca*—pidgin Arabic on Zande, up-country Swahili on Kenya schemes, kitchen kaffir in Rhodesia—languages that were native neither to them nor to the settlers, and which tended to be learnt and used by expatriates only in the imperative tense.[2] Yet others, like the managerial staff on the Niger Agricultural Project, had to deal with settlers entirely through interpreters.[3] On occasion there have been total breakdowns of communication and understanding. On one Uganda Group Farm, for instance, all the farmers suddenly declined to pick the cotton crop without the

---

[1] One British group farm manager on a two-year contract in Uganda said that he had been advised in the Department of Technical Co-operation of the British Government not to bother to learn the local language (Interview, 1965).

[2] A survey of the vocabulary range of managers could be revealing. It might show that meanings such as 'advise', 'suggest', 'recommend' are all subsumed into the word meaning 'command'.

[3] Baldwin, op. cit., p. 194.

Manager having any idea even of what was at issue.[1] Misunderstanding could even reach a pitch at which it threatened the future of a scheme. On one irrigation project in Kenya, an expatriate manager holding a meeting with tenants in Swahili gathered the impression that they were asking for the scheme to be closed down, and took immediate action on that basis. Subsequent investigation showed, however, that during a discussion of damage to crops by gazelle the tenants had merely asked for the scheme to be enclosed with a fence. The same Swahili word 'funga' could have both meanings. There can have been few incidents as serious as this, but the cumulative effects of innumerable minor misunderstandings and frustrations has often taken a heavy toll of managers' patience and settlers' satisfaction.

With post-independence schemes, the position improved with Africanization, but was sometimes worsened with the arrival of technical assistance staff. African senior staff usually either spoke the *lingua franca* fluently, like the A.A.O. in charge of the Tebere section on Mwea in 1966, or knew the local dialect, like the A.A.O. in charge of the Thiba section. African staff also had a better understanding of the social problems of settlers than their expatriate predecessors. On the other hand, technical assistance experts from assorted countries could have problems of adjusting to two intermingled cultures, African and British, and also of trying to communicate in English, of which they sometimes had only limited knowledge.

Cultural differences tend to reinforce a separation of staff and settlers which is anyway latent in the settlement situation. The most effective communication is direct, that is, face to face. Nyakashaka presents an exceptional case of direct communication in that the expatriate manager spoke the local language fluently, entertained the settlers in his home, conducted church services for them, supervised them individually on their plots, and advised them on their personal problems. But such close contact is rare. In other schemes, direct management-settler communication has been less frequent and has been more common in the early stages when staff are most numerous and training and instructing settlers are important staff activities. The individual staff-settler encounters which occurred in the early days of Gezira, and the public meetings and general assemblies of settlers which took place on Mwea in 1957, are examples. In the later stages of schemes, however, routine and bureaucracy reduce direct contact. Staff and settlers are anyway usually housed in separate residential areas and informal contacts outside the work situation are unusual. Expatriate staff, outside their work, often either withdraw into solitude, or engage in intense social activity with other

[1] Simon Charsley, 'The Group Farm Scheme in Uganda: a Case Study from Bunyoro', op. cit., pp. 14–15.

expatriates. Nor is this separation limited to senior staff or to expatriates. On Mwea senior African staff with experience of field agricultural work outside the Scheme are conscious of a more distant and formal relationship between Settlement staff and tenants than between field agricultural staff and ordinary farmers. Moreover, even junior staff feel themselves distinct from tenants and are reluctant to fraternize closely with them, for instance by drinking together in bars.[1] Social life as well as cultures and roles hold staff and settlers apart.

Both from inclination and necessity, managerial staff have resorted to and made use of indirect communication: the inclination has been strengthened by the decline of paternal values and the dominance of the technocratic-disciplinary approach; the necessity has been provided by the demands of economic imperatives and the inertia of organization and routine. Indirect communications use three principal channels: paper, junior staff, and settlers themselves.

Paper is a common means of communication, particularly on large and complex schemes, and usually has the effect of enhancing the authority of the management and increasing the sense of subordination of the settler. Mannoni has pointed out a certain 'liberation of the individual' resulting from the possession of government papers, providing a sense of immunity from hostile magic, and a sense of freedom of movement.[2] In a few cases this may have been true of settlement schemes. Certainly, the 'potential co-operators' of Ol Kalou saw in their photographs and identity cards a symbol of their right to the privileges of the scheme.[3] But far more often, paper has conveyed orders or rules. The written word carried high authority with illiterate or partially literate people: the impact of the vernacular bibles at the time of the Reformation, the implications of the West African proverb 'Book no lie', and the effect upon settlers of written contract agreements, are similar examples of the same phenomenon. In applying his thumbprint to a legal contract which he cannot read, a settler carries out an act of faith and abandonment, not fully sure to what he is committing himself. The contract, and other documents like loan agreements, warning letters, and summonses, may be intended as one-way authoritative messages from management to settler, and their effect can be accentuated by laborious and inaccurate translations made for settlers by friends or schoolboys. When settlers themselves try to communicate with management through paper, the written word is much less effective. Management reactions

---

[1] Interviews (1966).

[2] Mannoni, op. cit., pp. 150–151.

[3] The identity card could be withdrawn by the Leaders' Meeting, but by mid-1966 no such withdrawal had taken place (interview with General Manager, Ol Kalou Salient, 1966).

to letters are partly a matter of personal style, but on Mwea, where most tenants' letters have been applications, replies have often been stereotyped. Whether the written word is used by staff or by settler, thus, its effect is to maintain or increase distance, and to add to the authority of staff.

The importance of communication through junior staff varies with the ineffectiveness of other channels of communication. Descriptions of schemes where there has been direct communication, such as Nyakashaka and Upper Kitete, barely mention junior staff. But where they have had a near or complete monopoly of communication, they have acquired not only considerable power, but also an interest in maintaining their monopoly. The Azande Chiefs, for instance, acted as:

> a barrier between the people and the administration rather than as a channel of effective communication, for they depended on the people's ignorance of administrative regulations and always did their utmost to prevent direct communication between individuals and the administration.[1]

Junior staff on the Zande Scheme held considerable power in the allocation of plots, and misunderstanding reached the extreme of people believing that they were prohibited from growing food except on land set aside for a cotton rotation, an idea that was far from the official intention.[2] Similarly, on Sabi, Roder found that suggestions of the Land Development Officers often reached the cultivator as commands after passing through two levels of African intermediaries.[3] Communication on Mwea is rather more direct. Although much of the senior staff's information is derived from Field Assistants, there is a fair measure of personal contact between senior staff and tenants, so that junior staff cannot completely monopolize the issuing of orders or the passage of information or requests upwards.

Indirect communication also takes place through settlers themselves in accordance with a pre-existing or created authority structure. In mass settlements where whole communities are resettled, as with Volta and Khasm-el-Girba, traditional leaders may retain their authority and be used by management for a sort of indirect rule.[4] In specific settlements, and where there is no clearly existing authority system, managements may create settler agents: the Head

---

[1] Reining, op. cit., p. 36.
[2] Ibid., p. 120.
[3] Roder, op. cit., p. 172.
[4] Interviews in Volta Resettlement towns (1965) and Thornton and Wynn, op. cit., p. 13.

Cultivators on Mwea; the Samads on Gezira; the group leaders on Upper Kitete and on the Ol Kalou Salient; the village headmen on Kongwa. The relationship can be strengthened by a regular payment, such as the Shs.50 a month paid to Head Cultivators on Mwea, or there may be an unsatisfied desire for payment as on Kongwa.[1] In any case, traditional leaders and appointed agents are in positions similar to junior staff in which they have a clear personal interest in maintaining good relations with management and with the other settlers.

The most commonly found institution for indirect communication is the settler committee, typically presided over by the manager and attended by members of junior staff as well as settler representatives. On Mwea, the Tenants' Liaison Councils on each section, chaired by the A.A.O.s, and the Tenants' Advisory Committee for the whole Settlement, chaired by the Manager, are partly elected, partly appointed bodies,[2] which act as exchanges in which managerial explanation can be passed downwards, and tenant requests and suggestions passed upwards. The effectiveness of such committees as means of communication may decline with time, especially as leadership becomes stabilized or as they are captured by particular groups, especially of veteran, pioneer settlers. On Rwamkoma in Tanzania the original settlers' working committee continued for a time after the arrival of a second intake of settlers, who were not represented on it,[3] while at Ilora in Western Nigeria some pioneer settlers, in alliance with the management, secured a monopoly of the committee which was given a semblance of legitimacy by altering the village constitution. These pioneer settlers, whose position was resented by the younger settlers, invoked a tribal seniority principle and many came to perceive their committee positions in ascriptive terms.[4] In such instances, settler committees, like traditional leaders and appointed settler agents, serve as a barrier as much as a means to communication between management and the settlers as a whole.

Meetings of managers with settler representatives, particularly when they occur regularly, do, however, act as a source of authority and a focus for negotiation. A manager in committee can be not unlike a late medieval English king in parliament. In the king's (manager's) view he is answerable only to God (the parent settlement agency) and can even claim a divine right (delegated authority) to rule. In practice, though, any such ideas a manager may harbour are modified by

[1] Personal communication from Peter Rigby.
[2] The Tenants' Liaison Councils consist of the Head Cultivator and one elected tenant from each village, and are also attended by Field Assistants. The Tenants' Advisory Committee consists of five members elected by each of the Tenants' Liaison Councils and may be attended by some Field Assistants by invitation.
[3] Personal communication from Rodger Yeager.
[4] Okediji, op. cit., p. 307.

practical conditions which make it important that he gain and retain settler co-operation. A decision made in committee can sometimes carry greater weight, and appear more legitimate, than one made only by the manager, or only by the settlers. Here, however, there is great variety between schemes according to size, technical requirements and other conditions. The committees on Mwea are not regarded as in any way limiting the authority of the Management in technical matters, whereas on Tanzanian Pilot Village Settlements a classic process, similar to parliamentary acquisition of powers from a king, has been observed. Whatever the authority of such a committee, however, it constitutes a major adaptive and conciliatory organ for staff-settler differences. On at least one scheme in Tanzania the committee arbitrated between the technically-based wishes of the manager and the desires of the settlers.[1] Elsewhere, often in less explicit or less spectacular ways, they have provided managers with sounding boards and means of distributing information and orders, and settlers with a means of influencing management and passing up complaints, requests and suggestions.

Settler adaptations

That conflicts should arise between staff and settlers can be understood partly in terms of the stress and adaptations forced on settlers by settlement situations. Settlement involves uncertainty and a sense of insecurity. It can require a considerable act of faith on the part of settlers to submit to the risks, controls and changes entailed. It can face settlers with a need to make many major adaptations of which some of the most obvious have been: moving into unbroken country where there are unknown hazards such as spiritual dangers, wild game, and human and animal diseases, and where crops may not grow; accepting an unfamiliar agricultural process with uncertain rewards at the end and very heavy seasonal demands for labour; relying upon cash crop husbandry for money to buy food, without the security afforded by subsistence cultivation; submission to what appear to be, and sometimes have been, unnecessary controls and sanctions imposed by an arbitrary and distant authority; cultivating land without a sense of secure ownership, and with the threat of eviction recognized as real; accepting a large and imperfectly understood debt which may be felt as a great burden implying a surrender of freedom to government; living in an unfamiliar type of house placed in an unfamiliar spatial relationship to other houses; and even eating strange food. At the same time, even where the area of origin is close, or the settler has already been a migrant, settlement usually

---

[1] Personal communication from Rodger Yeager.

involves a disruption of social relations, a loss of economic and social assets, and isolation from the home society.

In addition, numerous other social problems commonly arise out of settlement. The authority of old men, traditional leaders, and priests may be diminished. Old men in the Kariba resettlement were expected to lose their socio-economic bargaining power since others who had previously depended upon them for land would clear new land for themselves.[1] Some chiefs in the Volta Resettlements lost authority when they moved off their land on to that of other chiefs. The fetish-priests of Volta and the shade-inheritors of Kariba were threatened by resettlement since their position was related to particular shrines or particular neighbourhoods that were to be flooded.[2] Elders at Mokwa feared that the Niger Agricultural Project would create a class of rich young men which would be a threat to their authority.[3] Again, family life may be seriously disrupted, if even only temporarily, as it was for polygamists on Volta who found themselves provided with only one room in which to accommodate their wives.

The lot of women is often worsened by settlement.[4] It is true that there have been cases where they have benefited, such as Gezira when the cotton boom of 1947–1951 so enhanced the incomes of tenants that, among the Arabs at least, it became regarded as shameful to compel a woman to go to the fields when labour might be employed instead;[5] and Upper Kitete, where mechanization of the whole wheat cycle released women from tasks they would otherwise have been expected to perform.[6] But such cases are exceptional. Settlement often deprives women of previous sources of income: the women of Kariba were expected to lose their cultivation rights to land,[7] and the women on the Niger Agricultural Project lacked raw materials for the occupations of spinning, dyeing and making palm oil which normally provided them with an income. They also had only limited opportunities for petty trading and marketing food crops.[8] Not only have

---

[1] Scudder, op. cit., pp. 144–145.
[2] Colson, op. cit., pp. 204–208.
[3] Baldwin, op. cit., p. 154.
[4] This topic is discussed for Tanzania in James Lewton-Brain, 'The Position of Women on Settlement Schemes in Tanzania' (mimeo), paper to Seminar on Rural Development, Syracuse University and University College, Dar es Salaam, 4 to 7 April 1966.
[5] Georges Brausch, 'Change and Continuity in the Gezira Region of the Sudan', *International Social Science Journal*, Vol. 16, No. 3, 1964, p. 352.
[6] Thomas (1966), op. cit., p. 5.
[7] Colson, op. cit., pp. 199–200, and Scudder, op. cit., pp. 69–70.
[8] Baldwin, op. cit., pp. 163–164. A similar situation limiting women's scope for petty trading has been noted by Kreinin in Western Nigerian Settlements. (Mordechai E. Kreinin, 'The Introduction of Israel's Land Settlement Plan to Nigeria', *Journal of Farm Economics*, Vol. 45, August 1963, p. 545).

women often lost their traditional sources of income, but their husbands have usually acquired a monopoly of new income: tenants' wives on Mwea frequently have not known what their husbands receive for their paddy, and bitter family quarrels have arisen from the husbands' new economic monopoly, similar to that of men working in towns.[1] Mechanization also usually places women at a disadvantage by carrying out heavy tasks, notably breaking and preparing land, which are often the traditional responsibility of men, but leaving to hand labour those tasks such as weeding normally carried out by women, the burden being aggravated by the additional acreages cultivated through mechanization. On Nachingwea the load of work on the women led to domestic upheaval and caused otherwise successful tenants to leave the scheme.[2] The disadvantages for women could, thus, affect the men indirectly, and the stability of the family and of settlement.

Yet another complex of social difficulties arises from what can be described as loss of society. When Rigby states that on Kongwa 'most of the organizational and ideological elements of Gogo society have had to be abandoned'[3] this may be taken as an example of a sort of social nakedness that has to be endured by settlers at least in the early stages of settlement.

Settlers employ a wide repertoire of defences against these problems and difficulties. The most universally recognized is an attempt to maintain links with the home environment except where, as with some refugees and those flooded out by dam lakes, it is not accessible or no longer exists. Visits for social and religious purposes have been common. The young settlers at Nyakashaka went out in search of wives. The settlers of Chesa initiated rituals in honour of their ancestors.[4] But with time, such visits and ceremonies become less frequent or important as social supports are developed within the settlements themselves.

A second response is to create new institutions and relationships to replace those that have been lost or abandoned. On Mwea, for instance, the village committee, originally an elected body, has become a semi-permanent judicial institution described by tenants as a *Kiama kia Itura*,[5] the traditional group of elders who initiate and judge in matters of custom, particularly marriage, and minor misdemeanours such as drunkenness. Similarly, on Upper Kitete a *Baraza*

[1] Lewton-Brain, op. cit., p. 3.
[2] Lord, op. cit., p. 90.
[3] Rigby, op. cit., p. 78.
[4] Woods, op. cit., p. 10.
[5] Kikuyu: village council. See Jomo Kenyatta, *Facing Mount Kenya, the Tribal Life of the Gikuyu*, Secker and Warburg, 1938, pp. 35 and 194.

175

la Wazee[1] was formed to deal with some customary judicial matters.[2] Settlement deprives settlers not only of traditional institutions, such as these were created to replace, but also often of close kinship relationships for which equally they need substitutes. On Kongwa, for instance, where kinship links were not as close as in the area of origin, different categories of kin were made to take on new functions.[3] Similarly on Chesa a new settler on arrival might seek out kinsmen and men from his own area, establish fictions of kinship with other settlers with the same name as himself, and form in-law relationships with persons belonging to the same clan as his wife. Mutual help would be pledged and labour often exchanged.[4,5] In such ways as these, settlers as it were reclothe themselves with social garments.

Quite apart from these social responses to the problems and disruptions of settlement, settlers also tend to adopt one of two polar attitudes to this situation: at one extreme a high degree of activity and self-help combined with individualism and independent attitudes; at the other a high degree of inactivity and apathy combined with dependent attitudes. Settlement appears to accentuate tendencies towards one or other of these extremes leading either to greater self-reliance or to greater dependence than in the pre-settlement condition. The fact that whole settlement schemes, and not just individual settlers, can be observed going one way or the other strongly suggests that it is not individual propensities which determine these responses as much as elements in the schemes themselves. Although many influences, not least the personalities and styles of managers, are relevant to any explanation, three factors—the method of selection, the provision of welfare, and the relationship between effort and reward—appear especially decisive.

In the first place, where settlers are 'selected' involuntarily by acts of god, government or war, all of which create refugee and evacuee situations, they easily acquire a sense of victimization which can lead to apathy and dependence. This is often aggravated by the loss of means of subsistence, the disorganization and unpredictability of an emergency situation, and reliance upon official food supplies.

---

[1] Swahili: 'Meeting of old men.'
[2] Thomas (1966), op. cit., p. 5.
[3] Rigby, op. cit., p. 59.
[4] Woods, op. cit., p. 10.
[5] The parallel is striking with Weingrod's observation of immigrants in the *moshavim*: 'In many cases, kinsmen who previously did not live near one another, or who never carried on co-operative activities, have joined together and now undertake joint projects', which suggests that this phenomenon may be general to settlement situations (Alex Weingrod, 'Administered Communities: Some Characteristics of New Immigrant Villages in Israel', *Economic Development and Cultural Change*, Vol. 11, No. 1, October 1962, p. 81).

Conversely, where selection is voluntary, the settler exercises and demonstrates his autonomy by applying for and accepting settlement, which he is then more inclined to regard positively as an opportunity rather than fatalistically as part of a sequence of disasters beyond his control.

Second, independent and dependent attitudes are closely related to the degree to which a scheme makes welfare provision for settlers, especially at the time of induction. On the one hand there are self-help schemes, such as Chesa, where settlers are left to clear their own land, build their own houses, and develop as largely independent smallholders. On the other there are welfare schemes, typified by the Western Nigerian Settlements, in which free food, monthly allowances, and high standards of housing and social services are provided or intended to be provided. Controlled migration schemes have approximated to the former type; *moshav*-inspired schemes to the latter. The most independent and active settlers are generally found where there is least 'scheme', and the most dependent and inactive where the greatest effort has been made to transform their way of life.

In considering this second factor, there is a continuity between the comprehensive paternalism of the colonial administrator, as exemplified in the Zande Scheme, and the enveloping benevolence of a would-be welfare state, as in the Volta River Resettlements, the Western Nigerian Settlements, and the Pilot Village Settlements of Tanzania. In both paternal and welfare schemes, the government assumed near-complete responsibility for providing what was regarded officially as desirable for the good of the settlers. Paternal settlements emphasized protection, and welfare settlements have emphasized the provision of social services, but the effects have been similar: inactivity, apathy, dependence, and a continual demand for more. In the paternal situation, these attempts can be understood as a form of patron-client relationship, in which a gift from the patron entitles and encourages an expectation of further gifts. In the welfare settlement situation, these attitudes can be understood as the rising demands of individuals who find themselves in contact with an apparently affluent government. In both conditions, settlers direct their energies and work off their frustrations not so much by productive activity as by political pressures upon authority. The more they are given, the more they demand, the less satisfied they are, and the more they blame the settlement agency for their real or imagined misfortunes. Settlers on Ilora were very dissatisfied with the facilities with which they had been provided,[1] and some settlers in Western Nigeria are said even to have blamed bad weather on the

---

[1] Okediji, op. cit., p. 304.

government.[1] A further factor increasing dependent attitudes among settlers in welfare schemes has been the incurrence of heavy debts to pay for services. In 1965 President Nyerere put it that:

> to burden the farm with very heavy debts at the outset, and at the same time, to make it appear that Government can provide all services, is not the best way of promoting activity. In future, we shall increasingly help by providing the economic services, leaving the development of the social services in the form of housing, etc., to the initiative and energies of the farmers as their work brings its return.[2]

These relationships were not a new discovery. In Nigeria, in particular, the same problems had arisen a decade earlier when the settlers on the Kontagora Scheme were assisted so generously that they came to demand such assistance as a right.[3] The converse was equally appreciated. Hunt wrote of Shendam that:

> It is felt very strongly that the principle of self-help should be the corner-stone on which all settlements are based, for without it no feeling of pride in possession will be engendered, and if once the feeling that the Native Administration and the Government will provide everything is allowed to become established, it is unlikely that the people can ever be left to stand on their own feet.[4]

A similar view was taken by the Preparatory Commission for the Volta River Project, which recommended that the people displaced by the Volta dam should resettle themselves. If they were resettled by government 'no matter what particular limitations or provisions might be made in this initial stage, constant pressure would inevitably be directed towards increasing the scope of operations. . . .'[5] The suggested self-help policy on Volta was, however, modified as the pressure of shortage of time combined with financial permissiveness to create precisely the situation which had been foreseen: a population demanding more and more assistance from government.

The third influence on settler attitudes, the relationship between effort and reward, is central to any understanding of settlement schemes. In non-scheme situations, the relationship is usually clear. A man works for an agreed wage, or he cultivates at his own risk

---

[1] Kreinin (*Journal of Farm Economics*), op. cit., p. 546.
[2] *President's Address to the National Assembly, Tuesday, 8th June 1965*, Mwanainchi Publishing House, Dar es Salaam, 1965, p. 15.
[3] Colonial Service Second Devonshire Course, 'Rural and Urban Settlement in the Colonies', op. cit., p. 14.
[4] Hunt, op. cit., p. 16.
[5] *The Volta River Project, Vol. I, Report of the Preparatory Commission*, H.M.S.O., London, 1956, pp. 51–52.

and benefits directly from what he grows. These may also be the conditions in the early stages of settlement if a monthly wage is paid, or in the later stages if the farming system is decentralized to settlers. In many schemes, however, returns to settlers have been controlled by the management or the settlement agency and have appeared to settlers to have been decided, as has indeed sometimes been the case, in an arbitrary manner by a distant authority.

Management and settler attitudes and behaviour, especially in relation to the economic process and returns, cannot intelligibly be explored separately beyond this point. Over time, in any given scheme, they develop as it were dialectically, in response to each other. They can be regarded as two teams which need each other in order to play a game, and which accept certain rules, but which are constantly adjusting and refining their tactics in response to one another.

Staff-settler interplay

Where schemes are continuing organizations, settlers' efforts to achieve security and satisfaction frequently conflict with the aims of staff. Such conflicts are particularly common over settlers' off-scheme interests, livestock, individual plots, and food supplies.

Off-scheme interests often provide a source of tension between managers and settlers. On Mwea, for instance, some tenants hedge their bets and improve their economic positions by starting off-scheme businesses, and although these businesses are not disapproved of in principle by the management, they are nevertheless liable to make successful operation of the rice process more difficult through divided commitments and absenteeism. Similarly, on Upper Kitete, when the settlement farmers rented land from neighbouring land-owners, this was almost bound to conflict with the management's desire for a regular labour turnout on the communal work of the scheme.[1] Off-scheme interests are a form of insurance and source of profit to settlers; but to managers they represent a threat to scheme production.

Livestock is another common focus of disagreement. Settlers frequently retain or buy cattle as their preferred form of investment and insurance but cattle are untidy and often in managers' views harmful to the main chance. On irrigation schemes, such as Mwea, cattle damage bunds through trampling. On tsetse barrier settlements, such as those in Uganda, they negate an original purpose of the scheme by providing stepping stones for the fly to cross the barrier. On the Kenya Million-Acre Scheme settlers resisted

[1] Thomas (1967), op. cit., p. 12.

179

official attempts to exclude native cattle where it was intended to maintain only high grade exotic animals.[1]

A further classic conflict involves individual plots where these are separate from what a management regards as the central economic process. During the early days of Rwamkoma, settlers were unable to spend the time they wished growing household food crops on their one-acre gardens because of the work required of them by the Manager on the cotton harvest which was intended as the source of the scheme's income.[2] On Mwea, the small areas of red soil available to tenants for rain cultivation are valued by them not just for the food crops they yield, which are unreliable, but also for the sense of security in carrying out a traditional activity which they provide. But this cultivation, while permitted by the Management, inevitably appears to it, like off-scheme businesses, as something of a distraction from the central activity of growing rice. In its simplest terms, this conflict is between settlers who place a high priority on assuring themselves of a food supply, and managers whose prime concern is growing and marketing a cash crop.

Where the cash crop is also a food crop, a major issue becomes delivery of the harvested crop by the settler to the scheme. A settler may be interested in retaining the crop for food; the management in obtaining high deliveries. This is liable to be a difficult matter for staff to control. On Mwea, the Management has permitted retention by tenants up to a limit of 12 bags of paddy per family. But by 1966 no effective system of checking had been devised, and in large families in particular more than 12 bags were probably quite often retained for food. Moreover, tenants were inclined to try to avoid Settlement charges by selling paddy on the black market outside the Settlement system, and a ready market existed partly because paddy was a food crop.

Many of these difficulties are softened by concession, compromise and concealment, particularly where settlers feel strongly about their interests or where managers do not see the conflicts as being dangerous to the central process. But where there are strong technical and economic imperatives, most notably mechanical or irrigational, management tends to be less flexible and to be preoccupied with securing effective settler compliance to strict requirements. This is usually sought first through persuasion, but where persuasion fails, formal disciplinary sanctions are often invoked. On Mwea, these are carefully graded. First, a tenant is warned verbally by a member of junior staff; then verbally by the A.A.O. in charge of the section; then in writing. If this fails, a court prosecution follows, and finally,

[1] de Wilde et al., op. cit., Vol. 2, p. 204.
[2] Yeager, op. cit., p. 43.

in cases of repeated misdemeanour or absenteeism, eviction is used. It is a sign of the staff's concern with production that over the five-year period 1961 to 1965, 94% of warning letters and 88% of prosecutions were for failure to maintain holdings, failure to comply with agricultural instructions, bad husbandry, water discipline, or absenteeism[1]—all faults which impeded agricultural efficiency. In the interests of production, the Management further reserved to itself the right to carry out by paid labour operations neglected by a tenant and then to charge him the cost, a proviso commonly found on other mechanized settlement schemes. Managers' reports from Mwea are revealing in the detailed breakdowns and analyses they provide of disciplinary action, including as they do details of types of offence, numbers of warning letters issued, prosecutions carried out for the different sections of the Settlement, average fines awarded, and percentage rates of prosecution and eviction. The concentration on discipline on Mwea may be exceptional, but it is usual on mechanized schemes to find sanctions used to secure compliance with agricultural requirements, with eviction as a last resort.

The sanctions available to a management are by no means limited to those with legal force. Indeed the comprehensiveness of the formal rules on Mwea is exceptional, and it is more usual to find that most of the penalties used on a scheme derive from the management's control of resources and of the central economic process. For instance, free food was withdrawn from recalcitrant refugees at Bushubi in Tanzania in an attempt to bring them into line.[2] More commonly, a factor in the production process can be denied to a settler. Water can be withheld on an irrigation scheme. On Urambo in Tanzania, Ilora in Nigeria, and the Kenya Million-Acre Scheme, refusal of loans or delay in their issue has been used as a sanction.[3] The Uganda Group Farms and Mwea are typical of mechanized schemes in that a settler who fails to carry out a required manual operation may be penalized through withholding a mechanical service. Moreover, where a management controls the payout there is always the possibility of informal fines through deductions before the settler receives his money.

For management, however, disciplinary action may be self-defeating. Sanctions and evictions tend to undermine settlers' security, confidence and commitment to a scheme, contrary to the official aim of settling them on it. As on Mwea, this may be countered through enabling settlers to pass on their holdings to

[1] M.I.S. Annual Reports 1961–1965.
[2] Yeld, op. cit., p. 12.
[3] Okediji, op. cit., p. 309; Ruthenberg (1964), op. cit., p. 85; and de Wilde et al., op. cit., Vol. 2, p. 204.

successors, and by formalizing procedures for evictions and allowing them only in certain well-defined cases. But even so, sanctions can still work against the main economic process since they drive settlers to hedge their bets with cattle, other businesses and subsistence crops, rather than rely too heavily on the scheme.

In contrast with sanctions a scheme's resources can be used by a management to provide incentives and to increase settler participation. On Gezira tenants considered to have performed their cotton operations well have been rewarded with permission to grow extra areas of lubia.[1] On Rwamkoma, scarce house frames were allotted by the Manager to those settlers who were thought to have shown special diligence.[2] On some schemes 'good' settlers have been given paid positions of responsibility, like the Head Cultivators on Mwea and the Samads on Gezira. 'Good' tenants on Mwea have been allowed to level and irrigate small extra fields in areas suitable for paddy rice but not included in the original construction programme. A further valuable privilege on Mwea, reserved for tenants whose recorded deliveries have been high, is seed-growing. In 1965, rice grown for seed for the following year's sowing on the Settlement was bought for Shs.54 a bag, compared with Shs.32 a bag for rice marketed in the normal way. Inducements such as these gain in effectiveness from the standardized economic base from which settlers usually start.

In manipulating sanctions and incentives, a management is forced to attempt to achieve a balance; sanctions based upon the central economic process tend to drive settlers away from full participation; incentives, conversely, draw them in to co-operate more fully. But penalties are generally easier to apply widely than rewards which are often from their very nature limited to relatively few individuals. Consequently, managements tend to rely more on sanctions and discipline than on positive inducements to secure required settler performance, and this tendency reinforces the division of staff and settlers into two separate teams.

The exercise of sanctions and controls is most convenient to managements, and most sensitive to settlers, at the point at which the returns to be paid to settlers for their produce are calculated. In complex and high capital schemes and schemes where marketing is co-operative, managements normally control payouts, so that it is relatively easy for deductions to be made. Such deductions can provide a powerful hold over settlers. Indeed, the calculation of returns is a central point in understanding most scheme systems.

Uncertainty in the settler's mind about the relationship between

[1] Shaw, op. cit., p. 12.
[2] Yeager, op. cit., p. 42.

production and reward is extremely common and disruptive, and has several origins. In the early stages of schemes a management itself may not know how returns are to be worked out. On the Niger Agricultural Project there was no intelligible system for dividing crop proceeds between management and settlers.[1] On Mwea there was at first uncertainty as to what water rate should be charged. On Upper Kitete in Tanzania, a battle was fought between settlers, manager and government, over the proportions of returns to be paid to settlers for their produce and to government for loan repayments. The position is complicated by the real power of many managers to secure remissions or subsidies for settlers on the grounds of unfavourable weather, the experimental nature of a scheme, or the need to maintain morale, so that settlers come to perceive that the rewards they receive depend upon the political pressures they bring to bear as well as on achievement in production.

A more lasting source of uncertainty is the opportunity for a management to make deductions before payout. These may be for many purposes. In paternal schemes, money was regarded by management as somehow inherently dangerous for settlers. It was felt proper that the management, in its wisdom, should use the money to the settlers' benefit, since they were unqualified to know what was best for themselves. On Daudawa, for example, returns were paid into a communal bank and then 'gradually doled out to the owners'. The Manager considered that, 'had it not been for the influence and persuasive qualities of the officers in charge, the greater part of (the money) would undoubtedly have been withdrawn the minute it became available'.[2] This control may not have affected the ultimate payout to the settler, but the possibility of retention of money in the bank by the Manager, or its use for communal purposes, was a potential threat. Indeed, the Manager wrote that he would like 'to impose an annual fee' which would cover among other things a water supply and anti-erosion work.[3] Similarly on the Zande Scheme most Europeans believed implicitly or explicitly that it was really desirable that the money income of the Azande should be low.[4] The control of returns on the part of the Equatoria Products Board was so complete and, to the Azande, so arbitrary, that they regarded their rewards as deriving from an unsatisfactory client-patron relationship with management as much as from their production of cotton.

In more recent schemes deductions have been made before payout

[1] Baldwin, op. cit., pp. 59–60.
[2] Taylor, op. cit., p. 73.
[3] Ibid., p. 72.
[4] Reining, op. cit., pp. 190–191.

o   183

for other purposes. Scheme charges for services and supplies, such as administration, mechanical cultivation, spraying and fertilizer, have been common. In addition, however, the concentration of money has been exploited by other organizations and interests. On Mwea the County Council cess is deducted at source by the Management and paid over in a single cheque each year. Court attachment orders have by law to be executed on tenant accounts, despite management reluctance to do this. Some contributions to worthy causes—the Thrift Society, the Harambee Secondary School, the hospital—have been permitted on presentation of a signed authority from a tenant. Although such contributions have been voluntary, the very fact that they are known to be so easy for tenants to make inevitably reduces their sense of freedom to refuse to make them. On other schemes contributions have sometimes been compulsory, and settlers have been captives of systems of which they have disapproved. On Ilora, for instance, mandatory monthly contributions to the Co-operative Thrift and Loans Society were unpopular and the fund was controlled by two settlers who used it in a discriminatory manner.[1] Again, co-operative societies for marketing can be used, as they have been on the Kenya Million-Acre Scheme, as a means of collecting loan repayments for the parent agency.[2] For various purposes such as these, whatever organization pays out to settlers tends to subject or expose them to financial commitments and obligations which they might prefer to ignore. In consequence, the value they place upon the central process of the scheme is reduced, and their search for other supports and for means of beating the system is stimulated.

It is precisely where there is a known and reliable system for relating individual performance to payout that cultivation activities are most efficiently and enthusiastically carried out by settlers. Of Urambo, Ruthenburg has written that the settlers

> are responsible for planting, harvesting and storage. . . . Each individual packs and sells his tobacco separately. In this way a direct relationship between the efforts and results of individual farmers is established. It proves to be the settlement's 'key to success'.[3]

It may be added that if such a system is to work effectively, the relationship must be intelligible to the farmer at all stages and he must also believe that he is receiving a fair price for his produce. On Mwea, the evolution and acceptance of routine and conventions

---

[1] Okediji, op. cit., p. 309.
[2] de Wilde et al., op. cit., Vol. 2, p. 200.
[3] Ruthenberg (1964), op. cit., p. 84.

had by 1966 reached a point at which the deductions made were predictable and generally understood, but they were none the less intensely scrutinized by the tenants.[1] The Settlement's system had been accepted by them as a fact of life, and their behaviour was adjusted to the assumption that little they could do would alter the principles upon which the Management made deductions before payout. It was not that their attitudes amounted to dependence or apathy, but rather that having largely accepted the system as unalterable, their individualism and desire for money drove them into working for higher production, into playing the Management's game.

The common interests of staff and settlers in higher production does not, however, obliterate the division of roles, identities and attitudes between the two teams, nor the misunderstandings and annoyances that may be experienced by both sides in their dealings with each other. Charsley has written of a meeting of Uganda Group Farm Managers that: 'considerable frustration was evident ... and much energy was devoted to considering ways and means of making life more difficult for Group Farm members.'[2] For their part, the members regarded the scheme as 'theirs', an imposition of a somewhat inscrutable government.[3] Similarly, on Mwea a member of the junior staff could say of the tenants that 'It is best to be tough as they work the opposite',[4] while the tenants describe the Management as 'the government', with overtones of distance and unwelcome authority.

A simple and obvious explanation of these divisions is to attribute them to the ethnic, cultural and attitudinal differences which staff and settlers have brought to schemes. Undoubtedly such differences have influenced the form that settlement institutions and processes have taken, but comparative evidence suggests that relationships of this sort, with communication problems, and conflicts and interplay between staff and settlers, arise out of settlement situations themselves wherever there is a relatively complex productive process.

The example of the Israeli *kibbutzim* is revealing. With their egalitarian ideology, democratic ethos, and stress on the value of communal decision-making, they might be expected to resist a division into two teams, the equivalent of staff and settlers. Yet out of communities consisting entirely of settlers, they have had to create managers: each *kibbutz* has its general manager, its treasurer,

---

[1] See Plate 8.
[2] Simon Charsley, 'The Group Farm Scheme in Uganda: a Case Study from Bunyoro', op. cit., pp. 16–17.
[3] Ibid., p. 26.
[4] Interview (1966).

and other specialized posts.[1] Over time developments towards formality, hierarchy, and specialization have been observed, and these have derived from the emphasis placed on production.[2] Similarly, the ideological settlements of the Ruvuma Development Association in Tanzania, while egalitarian and communal in style, have evolved and required a position of 'business adviser' to ensure that day-to-day economic decisions and negotiations are carried out, as well as an activist elite known as the Social and Economic Revolutionary Army.[3] Moreover, on the *kibbutzim*, a continual turnover in the holders of important managerial posts was held to be desirable, but the emphasis on the production and efficiency worked in the opposite direction. Rosenfeld has put it that:

> A young group is able to sacrifice efficiency for the sake of a principle it deems more noble. As the group increases in size, however, and as its investment in the enterprise grows larger, the risks entailed in maintaining a turnover of managers become too great. When this point is reached the managerial tasks are assigned only to those who form the small group of capable, energetic, trustworthy leader-managers.[4]

Furthermore, on Israeli collectives, as Rosenfeld has shown at length,[5] differentiation of roles led to social stratification into two classes which regarded each other with stereotyped attitudes. The 'rank-and-file' considered that the 'leader-managers' did not share their life and did not know what it was like. The leader-managers, on the other hand, often expressed their discouragement with the rank-and-file for their inadequate enthusiasm, their lack of public interest and participation, and their demands for a higher standard of living and less self-sacrifice.[6] The *kibbutzim*, indeed, despite their ideology, despite their initial and continuing sense of community, despite their recruitment of, in our terms, staff from settlers, had to develop organizations and social structures not unlike those of complex settlement schemes. There has, in fact, been a convergent evolution. On the *kibbutzim*, which began with a high degree of actual equality, change has been towards specialization and divergence of interests; on complex settlement schemes in anglophone

[1] Eva Rosenfeld, 'Institutional Change in Israeli Collectives', unpublished Ph.D. Thesis, Columbia University, 1952, pp. 60–61.
[2] Ivan Archie Vallier, 'Production Imperatives in Communal Systems: a Comparative Study with Special Reference to the Kibbutz Crisis', unpublished Ph.D. Thesis, Harvard, 1959, *passim*.
[3] Interviews with Ralph Ibbott and Ntimbanjayo Millinga (1966).
[4] Rosenfeld, op. cit., p. 72.
[5] Ibid., *passim*, but particularly pp. 71–77 and 90–92.
[6] Ibid., pp. 91–92.

Africa, which began with clear differences and inequalities between staff and settlers, change has been towards finding a common interest in the economic system. The progress of this convergence from different starting points has been largely determined by the requirements of the economic process. On settlement schemes it appears that in the long run it is not so much social relations which determine economic organization, as technical and economic imperatives which determine social relations.

CHAPTER 9

Organizational Ecology

Divided though they often are in the internal organization of a scheme, staff and settlers tend to combine in relation to their outside environment. They have a common interest in the survival and prosperity of their scheme, in securing services for it and in obtaining high returns for its produce. These of course depend partly upon the economic efficiency of the internal organization; but they also depend upon a scheme's external relations with other organizations, both in its local environment and more distantly with its parent agency. In order to examine these aspects of settlement schemes, it will be helpful first to consider the extent to which they acquire separate identities and can be examined ecologically, as organizations which adapt to and survive within an environment populated by other organizations.

Identity and survival

As a scheme is created it acquires territorial, social, organizational, and economic boundaries. In this process over time it gains a progressively clearer identity. Its geographical limits are surveyed, marked and recognized. The land within those limits is altered and used. Those who reside within the physical boundaries of a scheme and derive their livelihood from its land acquire particular status. This is acknowledged in a term, such as settler, tenant, co-operator or member, which distinguishes them from others. Through the development of institutions, as we have seen in Chapter 8, they adapt to the scheme situation and form a community or communities. Similarly, scheme staff become differentiated from non-scheme staff. In the formative stages of a scheme, the staff involved often do not 'belong' to the scheme, but are transient specialists such as surveyors, planners and builders. After settlement, however, the staff tend to be residential and to be fitted to the land in the pattern of their organization, controlled irrigation schemes providing an example of a hierarchy interlocked with a physical system. Like the settlers, they acquire territory over which they exercise forms of

188

authority. Although roles may become progressively more specialized, the lines of demarcation between settlers and staff may become fuzzy as settlers develop positions of authority, whether through leadership of other settlers, or through carrying out quasi-staff functions. Despite their differences, staff and settlers in these circumstances come to identify themselves with a scheme, and to perceive their common interests internally in production and externally in negotiating, as a political and administrative entity, for higher returns and benefits. The economic boundaries and the terms on which economic handovers take place become an important area of attention.

The emerging identity of settlement schemes can be illustrated subjectively in the ways in which they are spoken of and the ways in which people react to them. The territorial dimension is recognized in the syntax used by students of settlement after they have visited schemes: they speak of the situation 'on' rather than 'at' a scheme. Much of the identity of a scheme is most conveniently expressed in terms of ownership: land, roads, buildings, irrigation works, vehicles and crops may be spoken of as 'belonging to' a scheme. A man may be a settler 'on' a scheme, or an employee 'of' a scheme, the latter regarding the scheme as an organization, the employer. This is reinforced by the reactions of visitors to a scheme who ask 'Where is the scheme headquarters?', 'Where is the manager?', or 'Who is in charge?', reflecting an awareness that an organization exists, and a sense of unease, almost of trespass, a feeling that permission and approval may be needed to enter the territory without incurring an aggressive response from its occupant and owner.

In considering the identity and separateness of settlement schemes it is useful to distinguish four patterns of organization, according to whether a scheme *qua* scheme is transitory or persistent, and whether the territorial unit is extensive or isolated. In transitory situations, the agricultural and managerial system is such that there is little that can be described as 'scheme', and the area settled adopts the normal administrative pattern. In these transitory schemes, lasting identity has territorial and social but not organizational dimensions, in contrast with persistent schemes, where there is in addition an on-going organization. In extensive schemes the areas settled are so large that they create much of their own environment, and emerging identity is divided between an umbrella organization and the individual community and administrative unit. In isolated schemes, in contrast, the areas settled are separate, scattered and usually manager-sized. There are thus transitory extensive schemes, like the Sukumaland Development Scheme; transitory isolated schemes like some refugee settlements; persistent extensive schemes

189

like Gezira; and persistent isolated schemes, which are the most common and conspicuous types of settlement, including as they do the Volta Resettlements, the Western and Eastern Nigerian Settlements, Kenya's ALDEV schemes and irrigation schemes, the Pilot Village Settlements of Tanzania, and the Sabi Valley Irrigation Projects in Rhodesia. It is with these persistent isolated schemes, forming archipelagos of islands of settlement, that schemes appear most clearly as separate, bounded entities with an existence of their own, and which are principally considered here.

Island settlements lend themselves to ecological analysis by virtue of their isolation. Often they are physically remote: irrigation schemes in low hot valleys away from the main population centres (Perkerra and Tana in Kenya; the Sabi Projects in Rhodesia); resettlements from dam lakes which have taken place in hinterlands with poor communications (Volta); or miscellaneous schemes which are remote because only remote land has been available for settlement (Nyakashaka, Upper Kitete). In such projects communications are typically poor, with difficult road access, no telephone or telegraph facilities, and a slow postal service to the capital where the parent agency is located. In these circumstances, a young scheme organization can be regarded as an insecure intruder into an environment, endeavouring to establish a territorial base and to ensure its survival and prosperity. In pursuing these aims its external concerns are with its local human and organizational environment, and with its parent. These two aspects will be considered in turn.

Island schemes and their local environments

Staff and settlers on island schemes often show reserve and suspicion towards local people and organizations. Island settlements are frequently 'manager-size'. In circumstances of physical, social and organizational isolation, a manager's possessiveness easily asserts itself, and may take such forms as coolness towards visitors and local organizations, reluctance to improve an access road, and sometimes an almost pathological resentment of outside interference with settlers. For their part, settlers, particularly where of a different ethnic group or where they owe allegiance to a different tribal authority to that of the immediate environment, may resist assimilation: some tenants on Mwea at one time tried to refuse to accept the authority of the Wanguru African Court, and in Shendam some settlers from outside the host Native Authority area refused to pay taxes to it.[1] Far from always wishing to be integrated into

---

[1] Colonial Service Second Devonshire Course, 'Rural and Urban Settlement in the Colonies', op. cit., p. 13.

their surroundings, both staff and settlers are often jealous of their separate identity.

For their part, host populations and authorities usually resent the establishment of what they easily regard as a foreign colony on their soil. The belief that settlement schemes were a device for European acquisition of land was not confined to Mwea. It is also recorded for Sabi,[1] the Niger Agricultural Project,[2] and South Busoga.[3] More commonly, there has been hostility based on local ethnic differences, like the Gichugu and Ndia resentment of Kikuyu *ahoi*. Such feelings have been particularly evident where a chief's authority over land or people has been threatened by the introduction of foreigners who do not accept his authority, as occurred with Shendam,[4] some Volta Resettlements, and in at least one of the Sabi schemes.[5] Equally, politicians, local government representatives and officials, and the local officers of central government departments may view a scheme as an alien intrusion which does not concern them, a 'baby' of central government, a specially privileged area, the population and organization of which are not their responsibility.

The attitudes of both the staff and settlers on the one hand, and local people, the local government, and local officers of central government departments on the other, have, however, been ambivalent. Managers and settlers may fear domination by local authorities and organizations but nevertheless want the services they can provide. Similarly, the local people, local government and local departmental officers may resent a scheme but may also come, as the Gichugu and Ndia did with Mwea, to regard it as an opportunity. Chiefs, local governments, or departmental officers may lose authority through a scheme; or they may be able to increase their influence or usefulness by full participation in it or even by taking it over. But the general tendency, arising from this ambivalence of attitudes, coupled with uncertainty about roles and the newness of most schemes which have been studied, is for schemes to lack close contact with district level organizations, both administrative and political.

Administrative attitudes can be illustrated by the relations between government departments and Mwea. Departmental officers at the district level regard it as a special entity.[6] To the Medical Department it presents a professional challenge and interest because of the

[1] Roder, op. cit., p. 78.
[2] Baldwin, op. cit., pp. 28–29.
[3] Watts, op. cit., p. 19.
[4] Hunt, op. cit., p. 10 and Schwab, op. cit., pp. 19–20.
[5] Roder, op. cit., p. 173.
[6] The observations in this paragraph are based on interviews in 1965 and 1966 both on the Settlement and in the District Headquarters.

problems of bilharzia and malaria, and it receives special staff and special attention. To the Education Department it presents difficulties: teachers have been reluctant to serve on Mwea away from their homes and in a hot climate; and the Settlement's agricultural cycle prevents the collection of school fees at the normal time. To the Veterinary Department it presents an intractable situation, since communal grazing on the Settlement hinders any attempt to improve or protect the cattle. Little attention has therefore been paid to tenants' livestock. To the Department of Agriculture the Settlement is largely outside the limits of district concern. It is an area in which agricultural supervisory functions are carried out by the Management. Functionally, the territory is taken as being occupied. Thus, although the red soils within the Scheme are cultivated by tenants, Agricultural Assistants do not pay them any attention. Of these four perceptions of the Scheme, which may be expressed as a professional challenge (medical), a nuisance (educational), an intractable problem (veterinary), an already occupied territory (agriculture), only the first has led to special attention, while the latter three have been associated with a lower level of service than for non-scheme areas. A similar isolation of settlement schemes from regular government services has been noticed in Tanzania.[1]

Island schemes also tend to be isolated politically. On Upper Kitete a visiting Minister was distressed to discover that the settlers not only had no branch of TANU, but could not even sing the TANU song.[2] Kongwa had no village development committee at a time when they had been set up elsewhere in Tanzania.[3] The extent to which KANU in Thiba Location, which included Mwea, was an off-scheme organization was demonstrated by the fact that neither of the two KANU nominees elected unopposed in the 1963 local government elections was a tenant. KANU was weaker on the Settlement than off, partly, as a tenant enigmatically expressed it, 'because of the Scheme'.[4] Several reasons could be put forward to explain this statement and attempt to account for a scheme's political isolation: ethnic differences between scheme and non-scheme; the recentness of settlement, on the assumption that integration would occur over time; and a divergence of political interests between settlers and non-settlers. But the most cogent is that a scheme provides a political framework and system which focuses on the management as government. A settler committee which deals direct with a manager more closely affects the well-being of a settler

[1] Personal communications from members of the Syracuse University Team.
[2] Thomas (1966) op. cit., pp. 1-20
[3] Personal communication from Dr Peter Rigby.
[4] Interview (1966).

than a representative body at district or national level. The interests of settlers are better served within a scheme through 'political' activity directed towards the management, or outside a scheme through acting as a coherent scheme or settler interest group, than through generalized and relatively undifferentiated political activity within a wider party framework.

It is common to find that political matters are mediated for a scheme by an organization with administrative and political competence. Generalization is difficult, but the experience of Mwea may not be unusual. The organization in the environment most likely to attempt a take-over of a scheme (the Administration in the case of Mwea) may be converted into a buffer and broker between the scheme and other local organizations. It may be that this type of function would normally be performed by that organization which combines some or all of the functions of land negotiation, settler selection, settlement, and settler discipline and eviction, where this is not the management: TANU in the case of some schemes in Tanzania; the Social Welfare Department with Volta; the Advisory Committee and its sub-committee together with the Administrative Officer, in the case of the Niger Agricultural Project.[1] Such an organization may be of survival value to a scheme as a mediator and protector whatever the intentions of management and parent agency, whether expansion, maintenance of the *status quo*, or withdrawal.

In relation to their local environments, scheme organizations have used three principal forms of adaptation, which can be illustrated from the history of Mwea. In the first place a challenge or change may be ignored or rejected, with a consequent maintenance of autonomy with a degree of risk. The rejection of successive bids to take over Mwea are examples. Second, the response may be adoption, the incorporation of the external element into the scheme system. The progressive acceptance of Gichugu and Ndia as settlers, and the acceptance of local government representatives on the Local Committee are examples. Some loss of autonomy is compensated for by a gain in security. This is a very common response, and can be found in the advisory bodies often created for isolated schemes. Third, bargaining may lead to reciprocity, to an exchange of services in a symbiotic relationship. This could occur as economic bargaining, as in the argument that the rice cess levied by Embu A.D.C. would be in exchange for the A.D.C.'s acceptance of some financial responsibilities for the special medical services required by the Scheme. Or it could be less explicit, as for instance with housing loans. The A.D.C. (later the County Council), with its more flexible

[1] Baldwin, op. cit., pp. 19–21 and 26–29.

193

accounting regulations, assisted the Settlement by holding bank loans for settler housing advances in a suspense account. This arrangement benefited both the Management, which could not open such an account, and the A.D.C. in that it added to its links with the Settlement from which it derived considerable revenue in exchange for otherwise slight administrative effort.

Both adoption and bargaining as adaptations are second line measures for survival, for both depend upon a scheme having something to offer; in adoption, mainly power or participation; in bargaining, economic resources. But this distinction between power and economic resources is somewhat artificial: for power and participation are only meaningful if there is an organization or a community in respect of which power may be exercised or participation enjoyed; and whether there is an organization or a community depends to a great extent upon the viability of its economic system. Without at the very least a food supply, and usually a prospect of a cash income that compares favourably with opportunities elsewhere, settlers will leave. Without salaries, whether paid by a central agency or internally financed by a scheme, staff will leave. At the threshhold between survival and extinction for a scheme there are of course social factors operating, but it is the economic system that is most easily manipulated to ensure the continued participation of staff and settlers. Above that threshold, it is primarily the economic system that endows a scheme with resources and opportunities which make adoption attractive and bargaining possible.

In practice, however, the position is often complicated by the relationship of a scheme to its parent agency. For scheme survival usually depends not only upon local environmental factors and economic self-sufficiency but also upon parental protection, which in turn is supported by the rationales which can be used to justify a scheme.

Protection by parent and by rationale

The survival of settlement schemes is affected by the fact that the parent agencies, like their scheme offspring, are also intruders in environments already occupied by other organizations. These existing organizations, or more strictly the actors within them, tend to regard the newcomer as a threat, as a rival for functions, resources and status. First, in terms of functions, settlement (as an activity of government thought to justify a separate agency) appeared late in the sequence of creation of government organizations in Africa, in a period when the most obvious niches were already occupied. There were existing claimants to most or all of the functions of

194

settlement. Departments of Agriculture were responsible for all agricultural matters, Public Works Departments and Ministries of Works for government construction, Departments of Co-operation for co-operative organization, and the Administration for social and political affairs. All these, at a national level and below, were liable to resent and oppose the new animal trying to assert its right to perform their traditional functions. Second, the new agencies have been perceived as rivals for resources. Funds allocated to settlement might otherwise have been available to another department. More seriously, competition has sometimes been for very scarce trained agricultural staff. In both Kenya, where the Department of Settlement responsible for the Million-Acre Scheme sought to recruit from the Department of Agriculture, and in Tanzania, where the Village Settlement Agency[1] sought to attract competent agriculturalists, the struggle for staff was a struggle for the very means of carrying out a programme at all. Funds could be obtained at a political level; but political pressures were far harder to bring to bear to secure release of staff from other departments. Third, in terms of status, wherever settlement is regarded as important politically it attracts attention and receives a priority which may be the envy of older departments. Indeed, the very creation of a settlement agency may be regarded by other departments as a slight to their standing. In Tanzania, for example, the formation of the Village Settlement Agency to carry out the new 'transformation' approach to agriculture amounted to a vote of no confidence in the gradualism of the 'improvement' approach of the existing Department of Agriculture, and was deeply resented. Thus, although the behaviour of organizations cannot be encompassed or explained in a few sentences, these interrelated rivalries and resentments—over functions, resources and status—can be seen as some of the seeds of the hostility commonly shown by existing organizations towards new settlement agencies.

This hostility can be expressed in several ways. New agencies may be criticized, as ALDEV was, as the fifth wheel of the coach. If they fail to grow they can be smeared as ineffective; if they succeed in expanding they are maligned as efforts at empire-building. They may be attacked in more subtle ways. Special advisers or investigators can be appointed to assess a settlement programme, with the expecta-

---

[1] A detailed organizational history of the Tanzania Village Settlement Agency is given in Nikos Georgulas, 'Structure and Communication: a Study of the Tanganyika Settlement Agency', unpublished D.S.Sc. dissertation, Syracuse University, 1967; and a shorter analysis in Anthony H. Rweyemamu, 'Managing Planned Development: Tanzania's Experience', *Journal of Modern African Studies*, Vol. 4, No. 1, May 1966, pp. 10–14. The interpretations which follow owe much to these two accounts.

tion that their conclusions will be adverse. Officers of other departments can be appointed to controlling committees where their contributions may be negative, especially through the delays they can introduce into the making and executing of decisions. Settlement agencies can be denied the staff they seek, or indeed effective means to seek them, as happened with the Village Settlement Agency in Tanzania.[1] Or, if other departments are compelled or persuaded to second staff, they are in danger of receiving only low calibre personnel. But if a settlement organization does establish itself and is clearly not likely to be killed, another response to its presence is for another organization to attempt to capture it, to incorporate it, and so to acquire its assets of functions, resources and status.

To view settlement agencies in this way is, of course, to see only one of several aspects of their establishment and survival. Bureaucratic behaviour is not only to be understood in terms of the jungle. There may be considerable co-operation between departments. Moreover, criticisms levelled at settlement programmes are not necessarily merely rationalizations of hostility to a new or rival organization. This is especially the case with adverse economic evaluations. The Million-Acre Scheme in Kenya, the Western Nigeria Farm Settlement Programme, the Uganda Group Farms, and the Tanzanian Village Settlement Programme, have all been attacked variously by officers of Departments of Agriculture and of Ministries of Planning and of Finance on the persuasive grounds that they have been poor investments compared with alternatives. Such attacks as these, because difficult to refute, and because of the authority accorded to economic criteria, have been especially threatening to the existence and expansion of settlement agencies.

Faced with this hostility and these criticisms, settlement agencies have employed a wide repertoire of defences. Some fall within the categories already used to describe the responses of schemes themselves. Challenges may thus be ignored, or bargaining or adoption may occur. New settlement agencies may also rely heavily on a powerful patron in the form of another organization or a political body. The Department of Settlement in Kenya relied upon the support of both the British and Kenya Cabinets. The Village Settlement Agency relied on the Rural Settlement Commission, a high-powered political body, presided over by the Vice-President, and sometimes directly on the Vice-President himself. In addition to these defences two others, dispersal of hierarchy and proliferation of schemes, deserve attention.

It is in the interests of the agency, insecure as it may be in the capital city, to establish a dispersed hierarchy as soon as possible.

[1] Georgulas, op. cit., pp. 179 and 203.

One of the weaknesses of ALDEV in Kenya was that it lacked a geographical hierarchy and its schemes were effectively controlled by a territorial organization in the form of the Administration. In contrast, the Department of Settlement in Kenya did set up such a hierarchy in the field—with Area Settlement Controllers, Senior Settlement Officers, and Settlement Officers—and strengthened its autonomy *vis-à-vis* other departments through the lack of correspondence between its administrative areas and those of districts to which most other departments were fitted. In Tanzania the Village Settlement Agency was less successful. The Commissioner for Village Settlement, threatened as he was by a persistent takeover bid by the Ministry of Lands, Settlement and Water Development, sought to move his planning section away from the capital and to establish four regional headquarters, but was defeated by lack of staff, and the Agency was in due course incorporated in the Ministry.

Whether or not a dispersed hierarchy is achieved, proliferation of schemes on the ground is also a powerful support for an insecure parent agency. Once schemes have been established, economic criticisms become more difficult. Not only may there be substantial production and a possibility of showing an economic case for continuing to exist, but even if there is not, unfavourable evaluations become harder to make. Island schemes, by virtue of their scatter and isolation, are often difficult to visit. Those near the capital can be given special treatment and can become national showpieces for distinguished visitors, diverting attention from less economic projects elsewhere. Moreover once settlers have been established, a parent agency can muster a powerful range of arguments based upon commitment and obligation to the settlers, who would become refugees if a scheme were abandoned. In addition proliferation is a relatively easy form of defence for an agency, since there are usually political pressures for a wide distribution of schemes as forms of patronage to constituents or to particular areas. In short, then, a parent settlement agency with a dispersed territorial hierarchy, with scattered schemes, and with settlers, presents a difficult target for critics, and improves its own chances of survival and its capacity to protect its scheme offspring.

Protection can take many forms. The most common are direct or hidden subsidies. The free food provided for the Volta evacuees; the community facilities for the Western Nigerian settlements;[1] the United Kingdom Government's direct grants towards the Kenya Million-Acre Scheme; the substantial grant elements in the early

[1] Jerome C. Wells, 'The Israeli Moshav in Nigeria: an Estimate of Returns', *Journal of Farm Economics*, Vol. 48, No. 2, May 1966 (pp. 279–294), p. 280.

financing of ALDEV in Kenya;[1] the provision of capital free of interest to finance the Zande Scheme;[2] the shelving in the early 1960s of the idea that Mwea might repay its capital cost; the financing of the special bilharzia control measures on Sabi from the provincial budget, 'thereby avoiding direct charges against the projects'[3]— these are but a few examples of an extremely common phenomenon. In addition indirect subsidies may be provided in many forms: interest rates which are artificially low; government staff whose salaries are paid by other departments; international aid staff who are paid by their home governments; government services met by departmental votes; machinery, materials, and communications made available without charge to a scheme. Again, schemes, like infant industries, may receive fiscal protection. For example, the onions grown on the Perkerra Irrigation Scheme in Kenya could not compete with imported onions, nor indeed rain-grown Tanzanian onions, without a protective tariff. Then again, many schemes are launched in artificially favourable economic circumstances and can only be sustained through continued protection and subsidy. Some, like the Niger Agricultural Project and the Damongo Scheme, eventually cease. Many others become politically difficult to abandon and survive by a combination of institutional inertia and continued featherbedding. As infant industries they remain dependent infants. Without continued feeding by the parent and protection from competition they would die.

Such special assistance, whether in the form of subsidies or political or administrative support, can be justified and maintained by rationalizations of the purposes of schemes. Goals may be regarded, for settlement schemes as for other organizations, as dynamic variables,[4] and as providing a range of adaptive responses with high survival value. The rationales for settlement schemes can be considered as a continuum, with political, social and ideological goals and criteria at one pole, and economic goals and criteria at the other. In the life of a scheme the initial goals are usually political, social or ideological, but are pushed with time towards the economic pole, often under pressure of the application by other organizations of economic criteria to evaluations. The Sabi Irrigation Projects present an example of what can happen to official descriptions of the purposes of schemes. At first the Projects were defended as necessary to provide a source of grain in years of crop failure and food shortage, but when this had ceased to be an issue they were

[1] *African Land Development in Kenya, 1946–1962*, op. cit., p. 14.
[2] Reining, op. cit., p. 151.
[3] Roder, op. cit., p. 192.
[4] See James D. Thompson and William J. McEwen, 'Organizational Goals and Environment', in Etzioni, ed., *Complex Organizations*, op. cit., pp. 177–186.

justified as providing means for resettlement of Africans displaced by the Land Apportionment Act.[1] Thus, deprived of the support of one purpose, the Projects adapted by finding another. It was only much later in 1956 that economic criteria were applied, and it emerged that the Projects were subsidised.[2] The earlier humanitarian and political descriptions had been supportive; but the later economic description was hostile.

The Sabi Irrigation Projects were similar to many other settlement schemes which are frequently not viable economically in the full commercial sense, least of all in the early years of production. It is in the period when settlement has been carried out but production is not yet in full swing that schemes are most vulnerable, since with the settlers already on the land it is difficult to sustain the full force of the non-economic justifications, but it is also too early for an economic description to be anything but adverse, unless concerned with projected performance. In Nigeria and Tanzania especially, the poor economic record of settlements and the orthodoxy that economic development was an unquestionable good has presented schemes with a strong challenge. Lacking adequate political, social or economic justifications, the response has been to adopt defensive quasi-economic descriptions which may be located near the middle of the political-economic continuum. These resemble political and social rationales in asserting present or future economic benefits.

These quasi-economic arguments have taken three principal forms. In the first place schemes have been described as experimental. Research, as Belshaw has pointed out for Uganda settlement schemes, can be a protective description.[3] In the Western Region of Nigeria the stated purposes of settlement schemes were modified to include experimentation with new crops and with new types of organization.[4] In respect of the Tanzanian Pilot Village Settlements, an official statement put it that one of the aims of the programme was to gain experience.[5] The second quasi-economic justification has been an alleged demonstration effect. The Western Nigeria Development Plan for 1962–1968 asserted that Farm Settlements were '. . . intended to show that farming can be a profitable and attractive way of life. . . . The impact which these Farm Settlements will make on the overall productivity of farmers in the region has never been in doubt'.[6] Similarly in Tanzania it was argued that the Pilot Village Settle-

[1] Roder, op. cit., pp. 104–117 and 196–199.

[2] Ibid., pp. 135–138.

[3] Belshaw, op. cit., p. 5.

[4] Kreinin (*Journal of Farm Economics*), op. cit., p. 541.

[5] *Rural Settlement Planning*, op. cit., p. 2.

[6] *Western Nigeria Development Plan 1962–68*, Sessional Paper No. 8 of 1962 of the Western Nigeria Legislature, p. 19.

P
199

ments would set an example to neighbouring farmers and that the lessons learned would be applied over surrounding areas.[1] Although sociologists and economists disputed that there was likely to be any positive demonstration effect, the argument, by its imprecision and generality and the difficulty of verifying it, supported the schemes. The third quasi-economic argument has been that settlement schemes attract resources that would otherwise not be available. An instance of this is the argument put forward for Western Nigeria that Farm Settlements made it possible to increase the agricultural budget for other services. Kreinin implies that Nigerian agricultural officers argued that the government had been wearying of agricultural progress and shifting its attention to industrial development, but that:

> the settlement project has brought a change in that attitude by providing a political show-piece in agriculture. It has given the politician something tangible to show, not unlike an industrial complex, when election time approached . . . a modern settlement complex . . . is a visible achievement that can be used to get votes. Consequently it has made more rather than less money available to agriculture, and to the extension services in the process.[2]

A less unconvincing variant of this argument is that settlement schemes, and particularly irrigation schemes, by virtue of their identity and organization attract foreign capital which would not be available for any other sort of investment. It is of course possible that these three forms of argument could be justified on strictly economic grounds, though they have often appeared weak. The point here is that they could be introduced and paraded protectively to enable settlement schemes to survive. They are, however, used as a second-best, and tend to occur in circumstances in which more straightforward economic evaluations would be unfavourable.

Whether schemes are protected in these ways or not their life histories may be seen from one angle as struggles for economic autonomy, for freedom from debt repayments to a central agency and from external financial controls related to that debt. But such struggles, as a result of the original non-economic basis of a scheme, may be partially self-defeating. Mwea has provided a glimpse into the dilemma posed by repayment: the need for a manager not to allow a scheme to appear so profitable that capital repayment is required. This is not an isolated phenomenon. Returns and rewards to a

[1] Thomas (1966), op. cit., p. 8.
[2] Kreinin (*Israel and Africa*), op. cit., p. 61.

scheme, which means primarily to the settlers, are rarely directly related to the value of production. In the first place subsidy varies inversely with economic success. Second, successful schemes may be used to support less successful ones.[1] Third, increased returns may be absorbed through a change from social to commercial approaches to accounting. This last is illustrated by the Zande Scheme, where increased returns from higher cotton prices encouraged a shift from the provision of subsidies to increasingly commercial accounting, which limited the benefits passed on to the producers. When prices fell, the official rationale for the scheme reverted to its social purpose and subsidies were resumed.[2] The government parent which can be liberal in distributing subsidies in adversity thus becomes parsimonious when a scheme is economically successful. But when adversity returns, it may, as with Zande, revert to its original protective attitudes. Survival of the scheme is thus assisted, but settler satisfaction is not, unless settlers are ignorant of the true state of affairs. Even if they are ignorant, they may still perceive diminishing returns on higher prices or greater production or both. The dependent relationship of a scheme to its parent may thus improve the chances of scheme survival but it also acts as an inbuilt system to restrain incentives to settlers to increase production.

Nevertheless, in the long run the survival of most settlement schemes as identifiable organizations depends more upon their economic strength than upon parental support or protective rationalizations; similarly the survival of most parent organizations depends more upon their overall financial position than upon any other factor. In the short run, personalities, alliances, and passing political influences may be important. But sooner or later, unless a scheme or a programme has become financially sound, awkward economic questions are asked, and the most effective defence against them is evidence of a sound economy. In the very long term, the best assurance of survival is economic viability.

[1] This has been the case among the Israeli *kibbutzim*. See Eliyahu Kanovsky, *The Economy of the Israeli Kibbutz*, Harvard University Press, Cambridge, Mass., 1966, pp. 136 and 144.
[2] Conrad C. Reining, 'The History of Policy in the Zande Scheme', *Proceedings of the Minnesota Academy of Science*, Vol. 27, 1959, pp. 6–13.

Part IV
Analysis and Implications

CHAPTER 10

Phases, Activities and Organizations

In the discussion of staff and settlers, of internal scheme organization, and of the organizational ecology of schemes, detailed examination of changes over time has been deliberately withheld. Time is, however, a vital dimension for understanding any institution or organization, particularly one like a settlement scheme which is deliberately created and altered during its life.

Since there is not much relevant information about organizational changes in other schemes this analysis begins with a comparison of the two schemes, Mwea and Volta, which it has been possible to study historically, and then adds scattered evidence from other sources. The approach is to see what activities were carried out and by what organizations, and how they were interrelated. Activities are taken as central to understanding since they link the organizations and resources that are brought together to form a scheme. The treatment is selective in giving more attention to on-scheme and local activities, and less to those, such as policy-making, financial negotiation, and staff recruitment which are normally carried out at a settlement agency's headquarters.

Mwea and Volta: a comparison

The Volta River Resettlements in Ghana differ from the Mwea Irrigation Settlement in Kenya in many ways. Volta originated in the need to resettle people displaced by the lake formed behind the Volta dam, Mwea in the official desire to resettle landless Kikuyu displaced by the Emergency. The most formative years for Volta (from 1961 onwards) were after independence and much of the style of the operation was determined by Ghanaians. The most formative years for Mwea (1954–1963) were in the colonial period and the style of development was determined by expatriates. In scale Volta was far larger: most of the 80,000-odd people made homeless by the lake had to be resettled, compared with Mwea's 1,244 families by the end of 1960. Moreover the settlers for Mwea were, in the main, voluntarily recruited and came from continuing communities, while the settlers for Volta were 'compulsorily' displaced evacuees who constituted whole existing communities. Settlers on Mwea came from segmentary societies and could speak the same language; the communities resettled in Volta were more hierarchical and spoke many different languages. Mwea became one geographical entity; Volta was scattered in 52 dispersed residential and managerial units. Mwea was based upon irrigation, but the intention with Volta was to introduce co-operative mechanized agriculture. Organizationally too they were different. From the start the Volta operation was directed and co-ordinated by the staff of the Volta River Authority in Accra, without formal local participation, whereas until the formation of the National Irrigation Board Mwea was loosely controlled from Nairobi and was, for a time at least, strongly influenced by a local committee.

Despite these major differences there were similarities. Both were, of course, settlement schemes: they involved the controlled movement and settlement of population. In addition both were relatively high capital schemes and were intended to introduce complex forms of organization for production. Given the wide differences between them, whatever features they are found to have in common may prove to be general to other settlement schemes which also involve high capital and complex systems.

A conspicuous similarity of Mwea and Volta is that both involved the same activities in roughly the same sequence.[1] The activities of thought, survey, negotiation for land, planning, construction, selection, settlement, economic organization, and to a limited extent withdrawal are to be found in both schemes. Moreover, these activities can conveniently be described in terms of three phases each with its particular dominant activities and characteristics: first, pre-settlement with its political pressures and technical activities; second, settlement and organization with an emphasis on welfare and production; and third, withdrawal, in which the centre of attention is specialization and devolution. These three phases will be considered in turn.

Phase 1 Pre-settlement: political pressures and technical activities

The pre-settlement phase is typified by urgency, which varies with

---

[1] The understanding of activities on Volta comes mainly from *Volta Resettlement Symposium Papers*, op. cit., *passim*; *Volta River Authority Annual Report for the Year 1963*, and also for *1964* and *1965*, Guinea Press, Accra; Quarterly Reports of the Volta River Authority (mimeo), December 1962 to 31 March 1965, inclusive; files of the Resettlement Office of the Volta River Authority; and interviews with officials in Ghana (1965). Some Mwea and Volta findings are combined in Table 10.3 on p. 222.

the importance of a scheme's humanitarian and political rationale. Exceptionally, where the primary explicit purpose of a scheme is economic development, as with Gezira in Sudan and the Mubuku Irrigation Project in Uganda, technical considerations may dominate and largely determine the rate of implementation. But it is far more usual for the initial purposes of settlement schemes to be the solution of pressing human and political problems. As with Mwea there may be a landless population which poses a potential security threat. As with Volta and Khasm-el-Girba there may be a population of evacuees displaced according to timetables dictated by dam construction and a rising lake. In the case of refugee settlements there is usually a rootless population with inadequate means of livelihood. Or a drive behind settlement may be a desire for political support, for visible achievement and for power of patronage, as in post-independence Nigerian Settlements and the Group Farms in Uganda. But whatever the sources of urgency the impatience is the same. Though the circumstances were very different, the desires for quick results were similar when a senior expatriate administrator said of Mwea in 1956 that 'the pressure of an increasing number of landless detainees necessitates the progress of the Scheme being greatly accelerated',[1] and when a Nigerian politician, discussing the Western Nigerian settlement programme, exclaimed in 1961: 'Now when I look through all the agricultural projects in this Region, I find that what is being done is experiment, experiment, experiment, on and on. When shall we realise the results of these experiments?'[2] With such strong human and political drives behind the early stages of settlement, operations usually take place, as they did on Mwea, more quickly than is considered technically desirable.

One effect of the pressure for speed is a compressed overlap of activities.[3] On technical and practical grounds there is a necessary sequence of activities for a settlement scheme, especially from survey to planning and from planning to construction. Some overlap between these activities may be desirable, so that for any given scheme the optimum may be the equivalent (to take a physical comparison) of a partially collapsed telescope, with each cylinder representing an activity. Political forces exert a pressure to collapse the telescope while technical considerations resist. On Mwea the degree of collapsing and overlapping was regarded as excessive from a technical point of view. Equally, on Volta the programme went ahead far

---

[1] Memorandum, Special Commissioner, Central Province, to Minister for Agriculture, Animal Husbandry and Water Resources, 25 January 1956.
[2] Mr D. E. Okumagba, reported in the *Western Nigeria House of Assembly Debates*, 11 April 1961, column 504.
[3] For the overlap of activities on Mwea, see Table 10.2. on p. 221.

faster than the Agricolas would have wished had it not been for the unavoidable timing of resettlement.[1]

Some of the departmental differences of opinion which arise, though aggravated by speed, are primarily design-constructor or constructor-user problems. On Mwea designer-constructor problems only appeared in the early stages when the ALDEV Management were carrying out field preparation to designs made by the M.O.W. (late the M.O.W. was both designer and constructor); on Volta the town planners and the construction teams, however, were always distinct, and construction did not always take place according to the intentions of the planners. Constructor-user problems were more varied. On Mwea the Management at one time preferred the more convenient but more expensive arrangement of a separate inlet for each paddy field, against the M.O.W.s preference for only one inlet for each four fields; on Volta the Social Welfare workers would have preferred greater differentiation of housing, especially for chiefs, than the construction team was prepared or able to provide. On Mwea the M.O.W. handed over some fields that were not flat, to the distress of the Management which was faced with lower rice yields as a result; on Volta the constructors in one case handed over a town without completing the roads and drainage, which faced the Town Manager and occupants with problems of erosion.[2] On Mwea modifications of some channels were required to raise the water to a height that would enter the fields; on Volta modifications to the supports for the roofs of houses were required when a number blew off in storms.[3]

Other problems between technical departments are more directly a result of urgency. Three forms of difficulty can be identified. In the first place, communication and co-ordination can be ineffective under the stress of a tight timetable. The lack of exchange of information between the M.O.W. and the Management on Mwea had its counterpart on Volta in the difficulties of co-ordinating decision-making about site selection, and of ensuring an exchange of information between the town planners and those who were to carry out construction. A second type of problem can be termed overlap bleeding. As urgency provokes competition for limited resources, an activity that comes earlier in the formal sequence takes priority with the result that the later activities are bled of resources and fall behind schedule. On Mwea one reason why land preparation under the ALDEV Management was slow was that supervisory staff and labour

---

[1] M.S.O. Nicholas, 'Resettlement Agriculture', in *Volta Resettlement Symposium Papers*, op. cit. (pp. 86–99), p. 98.
[2] The town was New Mpamu.
[3] T. S. Johnson, 'Paper on Engineering Problems', ibid. (pp. 131–138), p. 138.

were required for the M.O.W.s work of canal construction, which was antecedent to the preparation of fields. On Volta overlap bleeding was conspicuous in the diversion of heavy machinery and labour from clearing agricultural land to clearing roads and building sites. The housing programme was consequently completed on time but the agricultural programme, for lack of cleared land, fell far behind. A third cluster of problems concerns delays and bottlenecks. The telescoping of activities can prevent the use of information which is not ready because an antecedent activity has not been completed. On Mwea this affected the soil survey, planning and construction.[1] On Volta the social survey information required by the town planners was often not available before the plans were prepared.[2] These three problems of lack of co-ordination, overlap bleeding, and delays and bottlenecks, thus aggravated the difficulties of the technical departments as they were subjected to administrative pressures for rapid execution of their functions; and the effects included interdepartmental friction, mistakes in development and delays in later activities.

The human purpose of these two schemes did not, however, only give rise to difficulties. In both cases high political priority meant easy money. In the early stages of Mwea abundant finance was available from both Emergency and ALDEV votes. With Volta the original estimate of £4 million for resettlement and compensation[3] was allowed to rise to £10 million.[4] This made it possible for most of the administratively complex problems of compensation to be blotted out through providing the evacuees with high cost housing which, in almost every case, was worth more than a family's compensation allowance. Financial permissiveness had the further advantage that the construction teams which were building the new towns could obtain supplies quickly without being strongly inhibited by cost. Indeed, some machinery for the programme was freighted to Ghana by air.

Despite these similarities, inter-departmental problems appear to have been more acute on Mwea than on Volta. There are many explanations for this contrast. Differences of attitude and administrative style between expatriates and Africans may have been important. The urgency and national prestige of Volta also helped to secure departmental support. Further, the form of organization adopted

---

[1] See p. 69.

[2] Laszlo Huszar, 'Resettlement Planning', ibid. (pp. 100–109), p. 107; and David Butcher, 'Social Survey', ibid. (pp. 28–37), p. 31.

[3] *The Volta River Project, Vol. 2, Appendices to the Report of the Preparatory Commission*, H.M.S.O., London, 1956, p. 135.

[4] *Volta River Authority, Annual Report for the Year 1965*, p. 8.

for Volta (an informal inter-departmental working party for the survey, site selection and planning stages; an *ad hoc* hierarchical organization for the construction operation; and a mobile team for the evacuation) were well adapted to the activities which had to be carried out.[1] It was perhaps even more important that roles were more clearly defined and accepted on Volta. These several factors go some way towards interpreting why there was less evidence of inter-departmental friction on Volta, especially over the boundaries of roles and over handovers, than on Mwea.

But beyond these explanations the contrasting departmental behaviour on Mwea and Volta can be understood in the light of three further inter-related factors affecting the stability and commitment of individuals and departments. These were the possession of a territorial base; permanence on a scheme; and the vertical range of activities performed by individuals or departments.

On Mwea the two main technical departments had physical bases. The M.O.W. at Gathigiriri and later Thiba, and the ALDEV Management at Kimbimbi, could each develop a sense of ownership with its staff, headquarters and land, and an absence of visible rivals. Both were also established on the Scheme with a degree of permanence, though once it was decided that the M.O.W. would not control water supply its presence was limited to the period of active construction or preparation for future construction, apart from occasional works of maintenance. Both achieved a deep vertical control of consecutive activities, the Management at first in carrying out every activity from survey to production on the Nguka swamp, the M.O.W. in colonizing the succession of activities from survey and planning to canal construction and eventually field preparation and structure maintenance. It seems reasonable to suggest that these factors contributed to the high degree of individual and departmental commitment to the Scheme.

On Volta, in contrast, the technical departments and individuals involved were rarely established on territorial bases, frequently on the move, and more specialized in their roles which were consequently shallower. The main departments were centrally-based in Accra and were dealing with geographically dispersed problems which discouraged the setting up of permanent or semi-permanent physical headquarters. They were essentially nomadic. The surveyors marking the limit of the lake were constantly migrating. The geologists carrying out water surveys never stayed long in any one place. The water drillers moved on as soon as their bore holes were established. The

---

[1] E. A. K. Kalitsi, 'Organization and Economics of Resettlement', *Volta Resettlement Symposium Papers*, op. cit., pp. 9–27, describes the organizations employed, especially for site selection.

planners were each responsible for a number of scattered towns, the sites of which they visited only intermittently. The construction teams were regarded as successful in so far as they finished each town quickly and moved on to the next. Too much of the work and control were based in Accra, residence on sites was too brief, and roles were too temporary and shallow, for powerful commitments or a sense of territory to develop among the surveyors, planners and constructors.

This set of explanations gains force from the fact that on Volta commitment was higher among the Social Welfare workers and the Agricolas who came into their own in the settlement and organization phase; for their work was centred in the settlements, they lived on the sites, and they had persisting roles in relation to the settlers.

Phase II Settlement and organization: welfare and production

With the act of settlement pressure for speed on political grounds loses its force; human problems cease to be as visible or important; financial permissiveness is replaced by financial stringency. The main aims, justifications and criteria for evaluation of a scheme shift from settlement itself towards self-sufficiency and independence of further assistance. Moreover, associated with these changes, activities and departments concerned with political and welfare matters are liable gradually to be displaced by activities and departments concerned with production and economic independence.

The parallel between Volta and Mwea in the performance of the activities directly involving local people is striking. In both cases the activities of local negotiation for land, 'selection', settlement, and welfare management after settlement were carried out by a single department: the Social Welfare Department on Volta, and the Administration on Mwea. On Volta land negotiation, 'selection' and settlement were difficult tasks. Obtaining land, according to the humane spirit in which the operation was to be carried out, required negotiation between the scattered communities which were to be aggregated, and possible hosts who might provide land. 'Selection' involved a detailed social survey and gave the people a chance to opt for official resettlement or to resettle themselves. This was followed by a long build-up to the evacuation and then a period of resettlement in strange surroundings and in larger communities than before. For these tasks the district administration might have been chosen; but the D.C.s were political appointees, lacked allegiance to the programme, and were unlikely to accept civil service co-ordination from V.R.A. Instead the Social Welfare Department, which already had

211

experience of resettlement elsewhere,[1] seconded Mass Education Assistants (M.E.A.s) and other staff to V.R.A. The position and authority of these M.E.A.s on Volta were somewhat different from those of the Administration on Mwea. Most of the M.E.A.s were relatively young and junior. They were also intruders into the local environment, not controllers of it. But they had powerful backing from V.R.A. in Accra, and a sense of mission, and claimed a monopoly of the right to act as brokers between the displaced people and the host communities. They became 'the main channel of communication between the Authority and the villages'.[2] Indeed, it was subsequently claimed that they acted as the only channel of communication in order that the people should not be confused by direct contact with the other departments working on the programme.[3] Not only did they execute the social survey, much of the site negotiation and the preparations for the move but, as the Managers of the new towns, they followed through their work through a continuing responsibility for the evacuees. The Town Development Committees which they, as Town Managers, took part in forming, provided them with a constituency, with an organizational as well as a territorial base.

The establishment of Town Managers in authoritative positions was easier because of an absence of rivals. The agricultural programme was continually falling behind and in many cases there was no agricultural presence in the town at the time of settlement. Whereas on Mwea the Agricolas were in occupation before the Administration, on Volta most of the Town Managers got to the resettlements first. Furthermore, with free food from the World Food Programme they distributed subsistence to the settlers after settlement and were thus able to dispense the type of benefit and reward—food and spending power—normally the prerogative of Agricolas. However, the real activities of production were neither assisted nor controlled by Town Managers. Agricultural staff, as they were posted in to settlements, formed co-operative committees and began the development of the surrounding land. The Town Manager controlled the town but the Agricola began to control parts of the surrounding countryside. At the same time Town Managers were uncertain about their future employment. Official concern about the need for more

---

[1] In the Tema Manhean operation, required to move the fishing village of Tema to make way for the port. See G. W. Amarteifio, D. A. P. Butcher, and David Whitham, *Tema Manhean, a Study of Resettlement*, Planning Research Studies Number Three, Ghana Universities Press, Accra, 1966, Part I, 'The Story of Tema Resettlement', by G. W. Amarteifio.

[2] *Annual Report of the Department of Social Welfare and Community Development, 1961*, Government Printing Department, Accra, 1963, p. 34.

[3] G. W. Amarteifio, 'Social Welfare', *Volta Resettlement Symposium Papers*, op. cit. (pp. 38–85), pp. 47–48.

free food, and the continuing economic dependence of the settlements, contributed to the shift of attention from town management to production. The change of emphasis was recognized in 1966 when the number of agricultural staff was increased and their salaries were raised above those of Town Managers.[1] As production became the dominant preoccupation, the scene was set for a takeover of roles by the Agricolas similar to that which occurred on Mwea from 1956 onwards. The more prolonged influence of the Town Managers on the Volta Resettlements compared with the Administration on Mwea can be understood in terms of their long established relationship with the evacuees, their early occupation of the towns, and the extended period of welfare and non-productive management resulting from delays in the agricultural programme. Despite these differences the lines of cleavage of roles between the Administration and the Management on Mwea, and between the Town Managers and the Agricolas on Volta, were similar, separating a cluster of social and welfare activities from a cluster of production activities. Since secure and effective settlement could only be achieved through economic independence for the settlers, the administrative or welfare department was in a weakening position.

Indeed, this division of roles and responsibilities and the tensions associated with it appears to be widespread. Of the Niger Agricultural Project, Baldwin observed:

> Few, if any, people seem to have recognized clearly that there were in fact two schemes running together in the guise of one. There were on the one hand the villages in which each settler had his house. ... On the other hand there was the mechanized farming scheme which it was intended to run as a commercial proposition. These two schemes were virtually independent. There was a natural tendency for the Government, through its administrative officers, to put the emphasis on the welfare of the settlers, whilst the CDC, through its staff employed on the Project, put the emphasis on the financial success of the farming.[2]

A similar split was noticed by Reining on the Zande Scheme, between the Administration which took a 'sociological' and welfare view of the Scheme, and the Staff of the Equatoria Products Board who took a commercial view.[3] It seems general for a department which may be termed 'the Administrators' (the Administration for Mwea and for most other Kenya schemes; the Social Welfare Department for Volta; the Administration for Gezira, the Niger Agricultural Project,

---

[1] Personal communication (1966) from Mrs Rowena Lawson.
[2] Baldwin, op. cit., p. 29.
[3] Reining, op. cit., pp. 153–159.

and Zande; TANU for some Tanzanian schemes) to gain or assert responsibility for matters political, concerning land, people, and welfare; and for difficulty to be experienced over the boundary between this set of roles and those concerning production which fall to the Agricolas. Welfare has generally proved a wasting asset, and the Administrators have been progressively undermined as settlers and Agricolas combine in production.

Phase III Withdrawal: specialization and devolution

With most settlement schemes a third phase can be distinguished in which the trend is for withdrawal of agency supervision and managerial control. This withdrawal takes two forms: specialization in which management and settlers, while concentrating on economic matters, call in specialized organizations to service a scheme, particularly in matters such as health, education, and roads; and devolution in which management hands functions over to settlers. The two are linked in that both are necessary if an agency is to withdraw from responsibility for a scheme without disruption. Although no clear boundary in time can be identified between the settlement and organization phase and the withdrawal phase, it is useful to discuss them separately since they arise from somewhat different pressures, are impeded by different obstacles, and are associated with different tensions and balances between organizations and groups.

Three principal sets of pressures for withdrawal can be identified. In the first place there are political and ideological grounds on which it is considered that settlers should manage their own affairs. With managers who are often expatriates and with the experience of political independence recent, it has been easy to transfer to settlement schemes the national model of political development in which alien rulers are removed and indigenous democratic institutions substituted. Further, the uncritical acceptance of co-operatives as being in the tradition of African socialism has in most countries given settler communities a convenient ideological lever for easing managers out of control. Almost without exception the settlement schemes of independence have incorporated or formed co-operatives, ranging in complexity from straightforward thrift societies as at first on Mwea, to communal production co-operatives as on Ol Kalou and Upper Kitete.[1] Although enthusiasm for co-operatives has diminished

---

[1] For a clear and useful survey of co-operatives on settlement schemes, see Edith H. Whetham, *Co-operation, Land Reform and Land Settlement: Report on a Survey in Kenya, Uganda, Sudan, Ghana, Nigeria and Iran*, The Plunkett Foundation for Co-operative Studies, London, 1968.

with the widespread experience of inefficiency and corruption that was almost inevitable with their rapid growth, the co-operative pattern of organization continues to pervade thinking about agricultural development and to make the handover of management to settler committees a widely accepted policy.

Second, settlers themselves often demand a greater say in scheme affairs, though their preoccupations change over time. Evidence from the minutes of meetings with management on Mwea suggest that soon after arrival on a scheme settlers are concerned primarily with subsistence, family matters, housing, security of tenure and discovering the rules that bear on their conduct. Later, interest shifts to finding ways of increasing production and returns from production. Only at this stage, on Mwea, did questions of representation become prominent, as also seems to have been the case on the Pilot Village Settlements in Tanzania and the Western Nigerian Farm Settlements. In parallel with these changing concerns a sequence of acquisition of authority by settler bodies can be discerned, starting with traditional matters such as adjudication in civil disputes and then expanding into scheme matters, organization of settler labour, the issue of credit, control of the central economic process including payout and even tractor operation, and negotiations with non-scheme organizations. In this process, settlers progressively take over authority and responsibility from management.

The third pressure for withdrawal comes from the settlement agency itself. After settlement has taken place, financial and economic criticism of programmes and shortages of funds often lead to cutbacks in expenditure. From one point of view specialization and devolution are a response to financial stringency, an attempt to achieve cheaper management. If a Ministry of Works can be persuaded to maintain a scheme's roads, a Department of Agriculture to supply agricultural services, a local authority to provide and maintain health and educational facilities, the agency (once responsible for all these) is saved expense. If settlers can be persuaded to manage their own affairs they cease to be a liability and commitment to the agency, and may even assist it. On the Kenya Million-Acre Scheme, for instance, the setting up of obligatory marketing co-operatives was intended to ensure loan repayments from settlers through inescapable deductions from co-operative payouts, saving costly administration by the Department of Settlement.

Given these ideological, settler and agency pressures for withdrawal, it is at first sight strange that in practice the process almost always takes longer than anticipated. On the Kenya Million-Acre Scheme the original intention was that management by the Department of Settlement would last only two and-a-half-years; but the period was later extended to five years, and it must be an open question whether the provision of special services for the settlement areas will ever come to an end. On Volta it was planned that supervision of the Resettlement Towns by the V.R.A. would cease at the end of 1965, but in 1967 it was still continuing without any signs of impending decisive withdrawal. In Tanzania, even after the official suspension of the Pilot Village Settlement Programme, most of the Village Settlements continued to exist and to retain managers. In Uganda, although Group Farm co-operative committees are expected to take over management, there has been little evidence of their doing so. In all these cases the transfer of responsibility and functions has been slower in practice than in the planners' theory.

There are many explanations for this protraction. Parent organizations have an interest in retaining control of 'their' schemes since to withdraw is organizational suicide: without schemes there is no reason to have an agency. Further, individual managers often find it difficult to abandon power, and their inclinations are to continue to manage at the very least the technical and economic aspects of schemes. Again, even when a scheme is patently uneconomic, officials and politicians are usually reluctant to take responsibility for signing its death warrant. Moreover, the fear that schemes will not be as well cared for by the indifferent or hostile organizations in their environments as they have been by the benevolent and sometimes affluent parent agency, is usually well justified. Ministries of Works and Departments of Agriculture often regard them as alien, and local authorities look on them as especially privileged communities which contribute little to revenue but which have been provided with high levels of services. These services, if taken over, would make heavy demands on limited resources which would be better used for the less fortunate communities which are, because longer established, also more influential politically.

But beyond these human and organizational brakes on withdrawal there can be persuasive technical and economic arguments for maintaining controls over settlement schemes. Where the agricultural process is complex, for instance with mechanization which requires a high level of managerial and technical skills, it can be argued that a project might collapse without continued supervision and support. Where, as most notably with the Kenya Million-Acre Scheme, settlers are required to repay heavy loans and where they show signs of not doing so unless subjected to sustained pressure, maintaining the settlement organization can plausibly be justified as a lesser, if more expensive, evil than writing off the loans. Where a complex irrigation system operates, as on Gezira, Mwea and Perkerra, the differentiation of staff roles is such that it becomes difficult to think in terms of

216

tenant technical management at all. In these ways both the motives and interests of those in the settlement organizations, and the objective situations with which they are dealing, combine to strengthen the status quo and to delay withdrawal.

Pressures for withdrawal affect the relationships between organizations and groups. Devolution may be marked by tension and negotiation between managers and settlers, with managers defending their authority on technical and financial grounds and settlers pressing for democratic independence and control. There have been exceptions: group farmers in Uganda who have been apathetic about co-operative organization and anxious to exploit the group farms with a minimum of commitment or effort; the settlers on Nyakashaka who asked the Manager to discipline their peers because they were reluctant to do so themselves;[1] and settlers on some Kenya Million-Acre Settlements who have requested that their Settlement Officers be retained. But the more general pattern of settler demands has been to acquire greater control. Normally, where a management provides mechanical or irrigational services it is not dislodged, for these call for special skills which settlers find it hard to muster, and require disciplined and carefully scheduled operation which is more easily achieved by a formal organization than by a community. It is over the later stages of the economic process that settlers more often assert their control. Processing, marketing and the payout are less absorbing than production to many managers, but come to be recognized by settlers as closely involved in the payments they receive for their crops. Thus on Mwea a degree of tenant influence and control has entered at the point of processing, through the investment of substantial tenant capital in the Settlement's rice mill. Elsewhere, as on the Kenya Million-Acre Scheme, the Tanzania Pilot Village Settlements and the Uganda Group Farms, settler co-operatives have to varying degrees taken up responsibility for collection, processing and sale of produce, and payment of proceeds to their members. They may also come to provide productive services, such as fertilizers, seeds, and even ploughing.

As attention shifts to devolution, processing, marketing and the payout, so an organizational niche is generated for professional co-operators. Much as, in the settlement and organization phase, Administrators concerned with welfare were displaced by Agricolas concerned with production, so now in turn Agricolas concerned with specialization and devolution may be displaced by Co-operators. Co-operators can claim to be an appropriate specialist organisation with a particular role in devolution, since democratic management

[1] Personal communication from Caroline Hutton.

217

is of the essence of the co-operative movement. The variations are many, and it is early to discuss this process, since many of the schemes about which information is available are only just approaching this stage. In Uganda, where the Group Farms are jointly supervised at the district level by the District Agricultural Officer and the District Co-operative Officer, competition and confusion have been mitigated by the lack of supervision accorded and the consequent autonomy of Group Farm Managers. In Kenya the Department of Settlement has received staff seconded from the Department of Co-operative Development to supervise settlement co-operatives, but in practice most of the assistance and advice received by co-operatives have come from Settlement Officers and Peace Corps workers, and it remains to be seen whether in the long run the Co-operators will dislodge the Settlement Officers. In Tanzania the Village Settlement Agency has sought to register its own co-operatives, but has had to compete with the Department of Agriculture and Co-operatives. At one time this led to a sharp cleavage on a scheme between a settlers' co-operative committee, registered with the Ministry of Agriculture, and expatriate V.S.A. staff working for the Ministry of Lands, Settlement and Water Development on the other. The rivalry reached a point at which the Co-operative was providing mechanical services with Jugoslav tractors, the staff were doing the same with Massey-Harris tractors, and servicing facilities were not being shared. This may be an extreme case of departmentalism but it is not isolated. In Western Nigeria co-operative organization was meant to be central to the Farm Settlements, but the fact that seven years after the Settlements had been initiated there were no active co-operatives managed by farmers has been attributed to bureaucratic difficulties between the Ministry of Agriculture, responsible for the settlement programme, and the Ministry of Trade and Industry, responsible for the government's co-operative programme.[1]

These various examples suggest that the withdrawal phase, like its predecessors, raises problems of departmentalism and of hand-overs: in specialization, between schemes and organizations in their environments; in devolution, between settlement agencies and managers on the one hand, and settler communities and departments of co-operation on the other. The problems differ in detail from those observed in the earlier phases, but are similar in reflecting changes in dominant activities and preoccupations and consequent shifts

---

[1] Carl K. Eicher, 'Reflections on Capital Intensive Moshav Farm Settlements in Southern Nigeria' (mimeo), paper presented to the A.D.C. Seminar on 'Co-operatives and Quasi-Co-operatives', University of Kentucky, 26–30 April 1967, pp. 8–9 and 15–16.

in the balances between the organizations which carry them out or claim a right and competence to carry them out.

A transformation model:[1] *the succession of activities and organizations*

From the description and analysis up to this point settlement schemes, like other organizations, can be seen as deliberate creations. They are brought about through the invasion of environments by people and resources. The people, resources and environment are linked by activities which produce change, and it is these activities and their sequence which are central to understanding what takes place. In complex high capital schemes the activities are relatively easy to observe since they are usually performed by specialized departments or individuals. On simpler schemes a manager may be responsible for most or all of the activities. Indeed, in a non-scheme form of settlement a man who spontaneously settles himself may perform all the activities on his own. But in all these cases the sequence of activities is similar, suggesting a model. The 'transformation model' used here separates out some of the main activities and describes some of their consequences, including the changing relationships of organizations which carry them out and the changing nature of the scheme.

For settlement schemes the main sequence of activities and their results can be summarized as shown in Table 10.1 on p. 220.

As these activities are carried out, so a scheme becomes defined as a separate, bounded entity. The boundaries emerge gradually but some activities have clearer results than others. Local negotiation leads to legal rights and definition of a geographical area. Construction asserts those rights through changing or building upon the land, and sometimes by marking the boundaries with fences or beacons. Settlement establishes social boundaries between settlers

[1] The term 'transformation model' is used to distinguish this approach from systems, organic, developmental and change-agent models, from all of which it differs in that a sequence of activities transforms an environment and creates a new entity. For systems analysis models, see David Easton, *A Framework for Political Analysis*, Prentice-Hall, New Jersey, 1965. For organic models, see Mason Haire, 'Biological Models and Empirical Histories of the Growth of Organizations' in Amitai and Eva Etzioni, *Social Change, Sources, Patterns and Consequences*, Basic Books, New York, 1964, pp. 362–375. For developmental and change-agent models see Robert Chin, 'The Utility of System Models and Developmental Models' in Jason L. Finkle and Richard W. Gable, eds., *Political Development and Social Change*, Wiley, New York, 1966, pp. 7–19. The transformation model crudely employed here could equally be used to analyse the creation of any development project or organization, with close parallels with, for example, a new town, a hospital, a school, or a business firm. As an analytical tool, it might repay more careful and systematic development than it receives here.

TABLE 10.1

Activities and changes

Serial number	New activity (general description)	What additional thing a scheme comes to resemble as a result of the activity
Pre-1	(Environmental change)	(Preconditions for an idea)
1	Thought	A proposal
2	Survey	Information
3	Local negotiation	A right to use land
4	Planning	Plans and estimates
5	Construction	A farm and housing
6	Settlement	A labour camp, an administered community
7	Economic organization:	
	Production	A farm in production
	Rationalization of economic activities	A commercial organization
8	Withdrawal:	
	Specialization	A specialized, interdependent organization
	Devolution	A self-regulating community
9	Growth:	
	(activities 1 to 8 again)	(as above, 1 to 8, expanding)

Note: The activities and their consequences are elaborated in Table 10.3 on p. 222.

and non-settlers. Production, except for internal scheme subsistence, requires and identifies handover points when produce is marketed. Throughout the growth of a scheme organizational boundaries emerge as staff are posted and integrated into a scheme system, and as interdependent relationships are formed with other organizations. These various boundaries may be subject to continuing change whether through expansion, specialization, devolution, or disintegration; but except in settlement which is barely scheme at all it is rare for some of these boundaries not to persist.

In practice these activities rarely take place in linear sequence. Instead they overlap. A general indication of the overlap of activities on Mwea is given in Table 10.2. In that case the degree of telescoping and the resulting bottlenecks and confusion may have been exceptional even for settlement schemes. But it is common around the period when settlement takes place for many activities, sometimes a bewildering number, to occur simultaneously. The situation is complicated when different geographical areas in the same scheme are at different stages of development, so that there is, as it were, a flow of the succession of activities over the land. In some cases the completion of an antecedent activity is vital before its successor

TABLE 10.2 *The Overlap of Activities on the Mwea Scheme, 1945–1961*

TIME BY YEARS

ACTIVITIES	1945	1946	1947	1948	1949	1950	1951	1952	1953	1954	1955	1956	1957	1958	1959	1960	1961
1. THOUGHT																	
2. SURVEY																	
3. LOCAL NEGOTIATION																	
4. PLANNING																	
5. CONSTRUCTION																	
6. SETTLEMENT																	
7. ECONOMIC ORGANISATION																	
8. WITHDRAWAL (Specialisation)																	

Notes: 1 The times at which activities are shown as starting and finishing are only approximate. Lighter and darker shading crudely indicate less and more intense activity.
2 Not all activities were continuous, although they are shown as continuous here. In particular, survey and local negotiation from 1949 to 1953 are shown as continuous but were probably sporadic.

TABLE 10.3

Phases, activities and organizations for a complex settlement scheme

A	B	C New activity (general description)	D New activities (specified)	E Individuals or groups carrying out the new activities	F Some common inter-group issues — Groups concerned	F Issues	G What additional things the Scheme comes to resemble as a result of the activities
	Activity	Pre-1 (Environmental change)	(Emerging perception of problem or opportunity)				Immediate pre-conditions for an idea
		1 Thought	Variable	Variable	Variable	Variable	A proposal
		2 Survey	Agricultural experiments	Agricolas — Investigators— Surveyors (various)	Administrators x Technical Departments (also in 4 and 5)	Speed	Information
			Surveys—land, soil, etc.		Surveyors x antecedent Surveyors / Designers x Surveyors	Promptness and quality of work	
PHASE I		3 Local negotiation	Identification and negotiation of land and other local rights	Administrators / Local right-holders	Administrators x Land right-holders	Terms of land concessions	A right to use land
		4 Planning	Agricultural layout planning / Design of structures / Budgeting	Planners (various)	Designers x Surveyors	Promptness and quality of work	Plans and estimates
					Users (Managers) x Designers	Convenience of operation	
					Constructors x Designers	Ease of construction	
		5 Construction	Land clearing / Layout of plans on ground / Field construction / Housing construction	Constructors	Designers x Constructors	Deviations from plan	A farm and housing
					Users (Managers) x Constructors	Quality and speed of	

	Activity		Administrators	Administrators x Potential settlers	Conditions of settlement	
	6 Settlement	Settler selection	Administrators	Administrators x Potential settlers	Conditions of settlement	A labour camp
PHASE II		Settler induction, organization, and training	Managers	Administrators x Managers	Control of settlers and settler welfare	An administered community
	7 Economic organization	Production—activities of the cultivation cycle	Managers and settlers	Managers x Settlers	Production activities	A farm in production
		Rationalization of economic activities	Managers	Settlers x Managers	Returns from crops	A commercial organization
PHASE III	8 Withdrawal	Specialization	Non-scheme specialist organizations	Managers and Settlers x other organizations	Provision of services	A specialized interdependent organization
			Co-operators	Managers x Co-operators	Supervision of settler co-operatives	
		Devolution	Settlers (taking over from Managers)	Settlers x Managers	Extent and timing of devolution	A self-regulating community
PHASE I-III	9 Growth (sometimes)	Activities 1 to 8 again	As above but Managers have wider control	Repeated as above	Repeated as above	As above, 1 to 8 expanding

Notes: This chart is based partly on the experience of the Mwea Irrigation Settlement and of the Volta River Resettlements. It represents a low level set of generalizations about what does happen, rather than about what necessarily ought to happen.

These activities are scheme-centred, and do not include many of those, such as major policy-formulation, financial negotiation, and staff recruitment, which are more likely to take place in an agency's off-scheme headquarters.

is started: a soil map should precede irrigation design; irrigation design should precede construction. But in any large scheme much can be gained by staggering the sequence geographically so that experience gained in areas developed earlier can be used in those which come later, an extension of the pilot project principle. The optimum degree and phasing of activity overlaps inevitably varies from scheme to scheme.

The sequence of activities provides a sequence of criteria by which a scheme's progress is judged. Here there may be a conflict between a manager's criteria and those of the parent agency. A manager, close to the ground, tends to judge progress according to the immediate task in hand: in construction he judges the work completed and its quality; in settlement, the morale of the settlers and evidence of their commitment; in economic organization, the scale of production and the returns received; in specialization, the extent to which other organizations can be persuaded to service the scheme; and so on. He commonly prefers to ignore financial considerations if he can. For the parent agency, however, these are usually vital. Up to the time of settlement both manager and agency may be united in their attitudes. After settlement at precisely the time when there is the most confusing overlap of activities at the scheme level, they tend to diverge. The manager's pre-occupations are settler presence, performance and satisfaction at just the time when the parent agency is liable to be coming under pressure from the economic ministries (the Ministry of Finance, the Treasury, the Ministry of Planning and Economic Development) and miscellaneous economists. In the capital city the earlier humanitarian or political rationale may carry little weight with those who demand quantification of achievement and financial viability. In Tanzania, for instance, assessments of the Pilot Village Settlements by government economists, scheme by scheme, were so adverse that managers were faced with major diplomatic tasks to ensure that what they considered to be adequate returns were paid to the settlers and not taken to cover capital and interest charges. As on other settlement schemes the managers' pre-occupations with settler satisfaction conflicted with the parent agency's interest in survival through financial respectability.

Perhaps the most significant aspect of the process of creation of a scheme is that it generates a succession of tasks to be performed which, in effect, provide niches which are occupied by organizations from the environment. These organizations may compete to perform the tasks, or they may have to be induced; but in either case the sequence of activities draws in a sequence of organizations. Table 10.3 lists some of the more obvious activities, the types of individual or group involved in performing them, and some of the common issues which arise between groups. The possible variations are legion, but in so far as the model is a valid description, it may be of use in anticipating needs and problems in a settlement programme, and in showing, in general terms, where consultation and co-ordination are important.

Implicit in the sequence of activities and of organizations carrying them out is a series of handovers to successor organizations and groups. As we have seen, these handovers can be difficult. In particular the handover of works by constructors, of settlers by administrators, and of managerial roles by managers often give rise to competition and conflict. The sequence of activities provides a framework for listing the principal handovers that can be expected with a complex high capital settlement scheme:

TABLE 10.4

Some handovers that may be anticipated in complex settlement schemes

Activity	What is handed over	Handed over	
		by	to
2 Survey	Information	Investigators	Designers
3 Local negotiation	Land rights	Local people	Administrators
4 Planning	Plans	Designers	Constructors
5 Construction	Works	Constructors	Users (Management)
6 Settlement	Settlers	Administrators	Managers
7 Economic organization	Produce	Managers	Marketers
8 Withdrawal: Specialization	Specialist roles	Managers	Specialist organizations
Devolution	Managerial roles	Managers	Co-operators, Settlers

If these handovers are recognized in advance as likely problems, it should be easier to reduce friction and difficulty through intelligent anticipation.

The extent to which the features described in this chapter can be identified in any scheme depends on many factors, not least the characteristics of the parent organization and the relationships of the scheme with its environment. They also depend on the internal nature of the scheme itself—its complexity, its stage of growth, its economic system, its form of organization. Up to this point the question of a typology of settlement schemes has been skirted with some circumspection. But now that staff and settlers, scheme systems, schemes and their environments, and changes in the nature of schemes over time have been considered, the raw material for a typology is largely exposed. In particular, seeing schemes as entities which change over time helps through directing attention away from the characteristics of schemes which are prominent in their early stages, and which are often used to label them, and towards their characteristics as mature organizations.

CHAPTER 11

Types of Settlement Scheme

In this chapter some of the observations and arguments presented so far are used in an attempt to derive and apply a classification for settlement schemes. Any set of categories, such as these, is justified only if it is useful. In some social science fields there are too many typologies for convenience, but this does not yet seem to be the case with agricultural development projects in general or settlement schemes in particular. None the less this attempt is made only with hesitation and reticence, and with a persisting fear of cumbering discussion with crude categories. For the limited purposes of this book the categories do help. Whether they have wider uses the reader will judge for himself. If they do not, at least they can be ignored.

In describing uniformities and differences between settlement schemes common usage seizes uncritically and inconsistently on their more obvious features. In their earlier stages schemes are frequently described by their initial purposes. Often the labels are provided by the origins and characteristics of the settlers. Thus there are schemes for school-leavers (Nyakashaka, Ilora), for the landless (Kenya Million-Acre High Density), for ex-servicemen (Shendam), for refugees (Bushubi), for urban unemployed (Kilombero), and for evacuees (Volta, Khasm-el-Girba). Sometimes the title comes from a special purpose not necessarily related to the settlers, as with tsetse control schemes (Kigumba, South Busoga). Where the economy of a scheme is dominated by a particular crop, that may be used as a definition. Mwea is often described as a rice scheme; other examples are cotton (Daudawa, Gezira, Rwamkoma, Zande), onions (Perkerra), sugar (Muhuroni), tea (Nyakashaka), tobacco (Kiwere, Urambo) and wheat (Ol Kalou, Upper Kitete). In these cases settlers have relied on other crops as well, but the one specified crop has been of central importance to the scheme's economy. Colloquially many other labels have been used: schemes have been described by size (Kenya *Million-Acre*), by title of settlement agency (*ALDEV* Schemes), by pattern of settlement (Tanzania *Village* Settlements,

Kenya Million-Acre *High Density* and *Low Density* Settlements), by status of land (Southern Rhodesia *Native Land Purchase Area* Schemes) or by imputed status of settler (the *Peasant* Farming Scheme in Zambia, the *Yeoman* Scheme in Kenya, and various *tenant*-farming schemes). None of these types of description has, however, been used across the board to cover all settlement schemes, and all of them are suspect of superficiality since they use criteria which are visible in the early stages of schemes, regardless of later developments.

More comprehensive attempts to classify settlement schemes have also been biased towards features obvious in the early stages of schemes. The separation of schemes into 'mass', 'incorporative', and 'specific'[1] and the employment of the expression 'controlled migration' are vulnerable to this criticism. Similarly, Apthorpe's tentative separation of three categories (sedentarization of nomadic populations, villagization of cultivators, and land settlement or resettlement) while useful in drawing attention to social origins and residence patterns, was not intended to penetrate deeply into the nature of on-going schemes.[2] Other criticisms can be levelled at the four classes of transitory extensive, transitory isolated, persistent extensive, and persistent isolated (used in Chapter 9[3]), since transitory and persistent are vague concepts, and in any case these categories were used *ad hoc* to separate out the persistent isolated schemes which most readily lent themselves to ecological analysis.

It might seem that descriptions of schemes according to the organization of production could provide a sounder basis for classification. There are, after all, what appear to be generic terms which are widely applied, such as group farming and co-operative settlement, both referring to forms of organization, and mechanical cultivation and irrigation, both referring to methods of production. But on examination these terms turn out to apply to a wide range of schemes. Group farming has been used to cover anything from alleged communal co-operation in Kenya around 1950, involving some land-sharing, to tractor hire services, with or without settlement, in Uganda in the mid-1960s. Co-operatives in settlement vary from marketing organizations largely run by settlement officials as on the Kenya Million-Acre Scheme, to communal systems encompassing land ownership, labour, production and the distribution of economic rewards as on Upper Kitete. Mechanical cultivation can range from individual settler enterprise in buying a tractor and hiring it out on a contract basis as on Chesa and some of the Kenya

---

[1] See pp. 150–152.
[2] Apthorpe, op. cit., p. 1. See p. 11.
[3] See pp. 189–190.

Million-Acre Settlements, to centrally co-ordinated and carefully phased operation of a management-controlled tractor fleet as on Muhuroni. Irrigation can be equally variable with at one extreme much of the water allocation and distribution in the hands of settlers, as on Kitivo, in Tanzania,[1] and at the other a closely disciplined organization controlled by a management responsible for all water movement, as on Mwea. In practice, thus, these terms are broad and imprecise.

In searching for a less superficial typology, a preliminary working classification suggested by Moris[2] provides a useful lead by drawing attention to the tendency for polar sets of characteristics to correlate. Moris describes five:

(i) Low new learning by settlers	High new learning by settlers
(ii) Low supervision	High supervision
(iii) Low specialization	High specialization
(iv) Low legal controls	High legal controls
(v) Farmer autonomous	Farmer dependent

Any scheme, it is implied, will tend to occupy a similar position between the poles on each of the dimensions. This approach provides a starting point since other characteristics also vary together in a similar polar manner. In Chapter 8[3] it was noted that size of managerial unit, length of organizational life and complexity of technical process tend to correlate. This can be expressed diagrammatically thus:

(vi) Small managerial unit	Large managerial unit
(vii) Short organizational life	Long organizational life
(viii) Simple technical process	Complex technical process

and can be added to Moris's five dimensions.

At this point much methodological and theoretical discussion might be introduced to argue a series of logical steps for deriving a typology. These would, however, probably be rationalizations for categories which have been arrived at intuitively. Suffice it to state here that three approaches or assumptions are involved: first, that settlement schemes are regarded as organizations; second, that they are not analysed in terms of their origins but in terms of the forms

---

[1] Personal communication from Stephen Sandford.
[2] J. R. Moris, 'The Evaluation of Settlement Scheme Performance, a Sociological Appraisal', *University of East Africa Social Science Conference Paper*, January 1967, No. 430.
[3] See p. 163.

into which they evolve; third, that the extent and duration of the controls exercised over the activities of settlers are a central principle in understanding and interpreting the polar characteristics already mentioned. Cause and effect statements are avoided but there are two sets of factors, one relatively immutable, one relatively adaptable, which appear to be powerful determinants of the form, extent and duration of controls, and of the patterns of organization.

In the first place the physical setting of a settlement scheme is relatively immutable. This is easily overlooked because it is so obvious and because land is usually altered in the early, often pre-settlement stages of a scheme. Yet the arrangement of land boundaries, of housing, of water supplies and especially of irrigation systems constitutes a fixed element to which the organization of a settlement scheme must adapt. Land boundaries strongly influence the lines of cleavage between scheme and non-scheme. Moreover, they are usually marked by limits of cultivation, fences, permanent beacons or earthworks, to provide a visible definition of a scheme. More important, perhaps, the attitudes of settlers and staff reinforce and sustain the boundaries over which they exercise control. Smallholders develop strong territorial sentiments. Their individualistic possessiveness is, indeed, so predictable that, like the physical framework of land layout to which it is fitted, it can be taken as a given factor wherever a sense of individual ownership is permitted.

Second, as has been argued scheme organizations converge on forms which are thought to maximize economic efficiency and returns. Among the many reasons for this are: the changing focus during the life of a scheme from human and political towards financial and economic criteria for evaluation; the professional attention of managers to production; the desires of settlers for high returns; the common interests of staff and settlers in scheme survival and prosperity; and the tendency of parent organizations to demand financial viability. These influences operate to fit the organization of a scheme to the requirements of the economic process.

Taking these two sets of factors (physical-territorial, and economic-organizational) and regarding controls as the key indicator, settlement schemes can be separated out into four more or less distinct groups along a rising scale of controls over individual activities by the management or community, starting with land tenure only, and then adding successively marketing, scheduled services, and a communal economy. Taking their most conspicuous features for labels, these types of schemes can be described respectively as individual holding, compulsory marketing, scheduled production and communal economy schemes.

230

Each type of scheme has a dominant imperative. In individual holding schemes it is the layout of the land and the possessive and territorial attitudes of the farmers. Lasting external controls are limited to land boundaries and to rules which apply elsewhere in non-settlement areas. In compulsory marketing schemes the imperative is primarily financial: the need to repay credit or a land purchase or development loan or to supply a processing factory with a crop. These are ensured through mandatory marketing and controls may extend to a regulation that a crop be grown. In scheduled production schemes the imperative is technical and derives from demands of efficient use of a centrally operated service, either mechanical, irrigational, or both. Controls over settlers include requirements that they should carry out operations according to a co-ordinated time-table. In communal economy schemes the imperatives are either technical, where communal farming has economic advantages, or ideological, where high value is placed on communal activities, or both. Labour and rewards are shared. Control over the lives of settlers has to be high, partly because the relationship between individual effort and reward is not clear, and this control usually extends into social as well as economic matters.

Controls increase cumulatively in moving from the simplest (individual holding) to the most complex (scheduled production and communal economy) schemes, and can be presented diagrammatically:

TABLE 11.1

Extent of controls by scheme types: Forms of control found

Types of scheme	Land boundaries planned and controlled	Central marketing required	Timing of settler production activities centrally determined	Labour and rewards shared
Individual holding	Yes	No	No	No
Compulsory marketing	Yes	Yes	No	No
Scheduled production	Yes	Yes	Yes	No
Communal economy	Yes	Yes	Yes	Yes

On this basis, the settlement schemes which have been considered most can be tentatively grouped as follows:

R   231

TABLE 11.2

A tentative allocation of settlement schemes to type categories

Individual holding schemes	Compulsory Marketing Schemes
Anchau (N)	Some ALDEV Schemes (K)
Chesa (R)	Daudawa (N)
Fra Fra (G)	Million-Acre Scheme (K)
Kariba (R and Z)	Nyakashaka (U)
Kigezi Resettlements (U)	Zande (S)
Kigumba (U)	
Shendam (N)	
South Busoga (U)	
Sukuma (T)	
Tanzania Refugees (T)	
Uganda Refugees (U)	
Volta (G) (some resettlements)	

Scheduled Production Schemes

a. Mechanical	b. Mechanized irrigation
Ilora (N)	Ahero (K)
[*Moshavim* (I)]	Gezira (S)
Muhuroni (K)	Khasm-el-Girba (S)
Nachingwea (T)	Mwea (K)
Niger Agricultural Project (N)	Perkerra (K)
Rwamkoma (T) (later stages)	
Urambo (T)	
Volta (G) (some resettlements)	

Communal Economy Schemes

[*Kibbutzim* (I)]
Ol Kalou (K)
Rwamkoma (T) (earlier stages)
Upper Kitete (T)

Notes:
1 Kiwere, Kongwa and the Sabi Irrigation Projects are not included, since the information available is inadequate for classification.
2 (G)=Ghana, (I)=Israel, (K)=Kenya, (N)=Nigeria, (R)=Rhodesia, (S)= Sudan, (T)=Tanzania, (U)=Uganda, (Z)=Zambia.
3 For some qualifications to this classification see pp. 234–235.

Each class of scheme is characterized and influenced by its physical form and economic system.[1] In individual holding schemes the central feature is the organization of holdings on a planned basis and the settlement of people upon them. Farm units are typically small and farm decisions are taken independently by settlers, who gain a strong sense of property in their land. Capital costs per settler

[1] Some of the characteristics of the four classes of scheme are presented diagrammatically in Table 11.3 on pp. 236–237.

are low and operating costs negligible or non-existent. Repayments, if any, for services rendered are on a once-and-for-all basis. Co-operation is *ad hoc*, resulting largely from settler initiative as needs are perceived. Official controls are slight and limited to land transactions and soil conservation rules which are similar to those applied outside the area affected by the scheme. Managerial style is advisory and sanctions are limited to normal social and legal procedures and prosecution and eviction are very rare. Official control of such schemes is brief and their ultimate condition is a merging with the normal administration of the surrounding area.

In compulsory marketing schemes the central feature is mandatory marketing and often cultivation of a certain crop or crops, at least for a period. This may be required to ensure repayments of credit or of land purchase or development loans, or to ensure a supply of crops to a processing factory, or some or all of these. Farm units are small and settlers develop a sense of property in land, but some cash crop farm decisions may be controlled by a management. Capital and operating costs are higher than on individual holding schemes. The organization for marketing the controlled crops is often a co-operative. Official supervision is limited to the production of the crops required for credit or repayments or for processing. Managerial style is both advisory and disciplinary, but prosecutions and evictions are rare. Some form of central organization persists to ensure seasonal credit, or regular supplies of a crop to a processing unit, or both, or may eventually cease if controls derive solely from a need to ensure loan repayments.

In scheduled production schemes the core feature is the organization of a centrally controlled technical service, mechanized or irrigational or both, upon which cash crop production on individual holdings depends. Settlers' land units are usually arranged for the convenience of operation of the central service, sometimes in separate strips, which limits the sense of property settlers acquire. Capital costs are high and in practice frequently subsidized or written off, and attempts are made to cover operating costs through deductions from the payout. In order to ensure those repayments marketing through a central organization, usually a co-operative, is obligatory. Official controls are considerable and are dictated largely by the need for a centrally-determined schedule for economic provision of the technical service. Managerial style is technocratic and commercial. Sanctions are commonly payout deductions to pay for services performed by the management when settlers fail to carry them out for themselves, the withholding of the central service, prosecution, and eviction. Official control of the scheme extends over a considerable period and if and when devolution to settlers occurs, the production imperatives of the central service continue to apply.

In communal economy schemes, in addition to a scheduled service, land and labour use are communal and rewards are shared. Even when they have small individually farmed plots settlers are slow to develop a sense of property in land. The main territorial unit that can be identified is the scheme as a whole. Capital costs are very high and operating costs substantial. Capital is in practice frequently written off to enable the scheme to survive, and the substantial operating costs are either subsidized or recouped by deductions before payout, or some combination of the two. The form of organization is often described as co-operative. Controls are considerable with a high degree of constraint upon settlers to perform activities at particular times. Managerial style is variable, with individual personality more significant than with other types of scheme, and is often ideological or inspirational. Sanctions are generally social and economic rather than legal, with formal eviction rare. Marketing, production activities and the allocation of settler labour are centrally determined. Ultimate withdrawal by the settlement agency may be intended but the scheme, if it survives, remains a special entity.

These generalizations describe what appears to happen, not what ought to happen. They are dogmatic for brevity, but tentative in fact. Undoubtedly exceptions can be found to them. Moreover they must be qualified at once in four respects.

In the first place, these categories are imprecise and not necessarily comprehensive. It is not always easy to see in which class to fit a prticular scheme, and there are probably schemes, not considered here, which do not fit any of the four classes and deserve a separate category or categories.

Second, the degree of control exercised in a scheme does not necessarily rise evenly through the sequence of types. The personality of a manager, the rules in force, the scarcity of a production factor or of a service, or technical requirements not directly related to marketing, mechanization or irrigation (such as the unpopular task of burning cotton stalks after harvest on Gezira) may be responsible for closer controls than would have been expected from the descriptions given above. Conversely a scheme may be unusually permissive, as it is with mechanical cultivation on Urambo in Tanzania, where farmers can either call in tractors or carry out cultivation activities themselves by hand.[1] Moreover, control is greater in the early stages of a scheme when a system is being worked out than in the later stages when it has become a routine.

[1] Personal communication from Kenneth Baer.

234

A third qualification arising out of these variations in control concerns settlers' attitudes. Dependent attitudes among settlers are strongest at first, and decrease as a system is learned and comes to be accepted as a fact of life. Dependence also derives from methods of selection, and the extent to which a settler becomes self-supporting. Despite these qualifications, the correlation between complexity of scheme system and dependent attitudes is important and its inclusion seems justified.

In the fourth place, any given scheme may change types during its life. A progression to greater complexity and control is possible, as when marketing co-operatives on Kenya Million-Acre Settlements start up mechanical services for their members. But it has been more common for there to be a slide in the other direction, from complexity towards simplicity. Rwamkoma in Tanzania was to have been based on communally cultivated cotton and would have been a communal economy scheme; but under pressure of the individualist desires of the settlers it became a scheduled production scheme with centrally co-ordinated mechanical services and individual cotton plots. The Volta Resettlements in Ghana were to have become co-operative mechanized farming units, with some individual cultivation and some central services; but on many of the settlements, when the mechanical programme was delayed, they took the form of informal individual holding schemes. Damongo, too, when it was abandoned, ceased to be a scheduled production scheme and survived in attenuated form as the Fra Fra individual holding scheme.

Indeed, the more complex a scheme the less stable it appears. Communal co-operative production is particularly difficult to sustain. To date it has been achieved with simple crops by the Ruvuma Development Association, through convinced leadership and strong ideological motivation, in groups which can be described properly as settlement but which are self-regulating and non-governmental and therefore not strictly 'scheme'. Elsewhere, it is striking that the two communal economy settlement schemes in East Africa which are known to have had some success (Upper Kitete in Tanzania, and Ol Kalou in Kenya) have been based on wheat and livestock both of which, in different ways, are susceptible to economies of scale. In contrast with cotton all the operations in the wheat cultivation cycle can conveniently be mechanized, and this in turn demands large land units. Similarly, there are labour economies in large herds of livestock: one man may as well look after fifty cattle as five. The implication is that if a communal system is desired on ideological grounds, it is prudent to choose a crop in which the advantage lies with scale, and which does not entail manual cultivation or harvesting of a sort which, as with cotton, may be most

235

Table 11.3

Some characteristics of four types of settlement scheme

	A Individual Holding	B Compulsory Marketing	C Scheduled production	D Communal Economy
1 Central feature of Scheme	Planned individual holdings	Mandatory marketing and often cultivation of a crop	Centrally scheduled production service provided for individual holdings	Communal land and labour use and sharing of rewards
2 Purposes of controls	Boundary and holding size maintenance	As in column A plus ensuring repayment of credit or loan, or supply of a crop for processing	As in columns A and B plus the efficient use of central mechanical and/or irrigation services	As in columns A, B, and C plus the maintenance of the communal system
3 Farm decisions	Decentralized	←– – – – – Mixed – – – – →		Centralized
4 Managerial unit size	Small	Small	Small	Large for marketing, central service, and labour use
		←– – Large for marketing	Large for marketing and central service	
5 Sense of ownership of land and resources	Individual settler	←– – – – – – – – – – – →		Management or communal
6 Capital cost per settler	Low	←– – – – – – – – – – – →		High
7 Operating costs per settler	Low	←– – – – – – – – – – – →		High
8 Repayments by settlers for services	Nil or once-and-for-all	Loan repayments (limited period) and/or credit and marketing service deductions (lasting)	Capital repayments (limited period, in practice often not required). Marketing and central service deductions (lasting)	Capital repayments (limited period, in practice often not required). Marketing, central service, and social welfare deductions (lasting)

236

	Ad hoc largely on settler initiative	Obligatory marketing, usually through co-operatives, to ensure credit and/or loan repayments, and/or crop supply to processing factory	Obligatory marketing usually through a co-operative, to ensure loan repayments and/or crop supply, and payment for central service	Obligatory co-operation integrated with communal system
9 Co-operative organization	*Ad hoc* largely on settler initiative	Obligatory marketing, usually through co-operatives, to ensure credit and/or loan repayments, and/or crop supply to processing factory	Obligatory marketing usually through a co-operative, to ensure loan repayments and/or crop supply, and payment for central service	Obligatory co-operation integrated with communal system
10 Technical imperatives and technical inter-dependence of settlers	Low	← – – – – – →	← – – – – – →	High
11 Continuing controls	Low Land transaction rules; Soil conservation rules	As in column A plus Marketing; Credit and loan repayment, crops grown (sometimes)	As in columns A and B plus Scheduling of farm activities	High; As in columns A, B and C plus comprehensive controls over production process
12 Management style	Advisory	← – – – →	Technocratic	Ideological and inspirational
13 Management sanctions / Special sanctions	Normal legal procedures; Prosecution and eviction very rare	← – – – →; Prosecution and eviction rare	Special legal rules; Prosecution, denial of service and eviction	Special rules (communally enforced); Social and economic sanctions with formal eviction rare
14 Settler attitudes	Secure; Autonomous	← – – – →	Insecure; Dependent	Mixed
15 Time span of agency control	Short with withdrawal	Medium-term withdrawal intended (usually delayed)	Long-term withdrawal often intended (usually delayed)	Ultimate withdrawal intended
16 Ultimate condition of scheme	Normal administration	← – – – →	← – – – →	Special entity

Note: The statements in this table should be taken as descriptions of observed general tendencies and not as laws or predictions. The dotted lines indicate a more or less continuous gradient between the two conditions.

efficient if an individual relationship with a piece of land and with the returns from that land is established.

At this point, it is intriguing to return to a question implicit in Chapter 1.[1] Is it useful to distinguish settlement from non-settlement schemes? In trying to answer this now it is at once evident that there is a parallel, non-settlement classification for forms of organization which do not involve the transfer of population but which do involve similar economic systems. These are:

TABLE 11.4

Types of settlement and non-settlement scheme

Settlement type	Non-settlement type	Examples of non-settlement type
Individual holding	Land tenure reorganization	Land consolidation (Kenya) Land centralization (Rhodesia)
Compulsory marketing	Co-operative marketing, rural credit, processing outgrowers	Cocoa marketing (Ghana) Credit schemes (Uganda) Kenya Tea Development Authority outgrowers (Kenya)
Scheduled production	Tractor hire Non-settlement irrigation	Some Uganda Group Farms Marakwet Irrigation (Kenya)
Communal economy	Government estate or plantation, state farm	Groundnut Scheme Ghana State Farms

Certainly the boundaries between settlement schemes and their non-settlement equivalents are not always distinct. Land consolidation in the Central Province of Kenya did involve local movements of population and so could have been described as a settlement operation. Some Uganda Group Farms include settlement and non-settlement under the same organizational framework. Nyakashaka markets tea for both settlers and nearby non-settlers. Eicher has written of Nigerian settlements that: 'Since the farmers are able to make only token managerial decisions on their farms and since the co-operative aspect of the farm settlement is not functioning, the Moshav farm settlements are in practice state farms rather than co-operative land settlements.'[2]

To the conundrum: when is a settlement scheme not a settlement scheme? there is no final answer. But in general settlement schemes involve closer controls and more compulsion than their non-settlement counterparts.

[1] See p. 10 ff.
[2] Eicher, op. cit., p. 9.

To pursue questions of definition and classification beyond this point would be unprofitable. What does matter is to expose and clarify the policy choices open to governments and to show what their implications are. Settlement schemes compete with non-settlement approaches for the same scarce resources, particularly finance, technical expertise, and managerial capacity. The question whether settlement or non-settlement approaches to agricultural development are to be preferred cannot be answered here, since only settlement schemes and not the alternatives have been examined; and anyway each comparison of choices is unique. In any case, quite frequently there is no alternative to settlement. It may be forced upon a government by circumstances. Nor, where it has to be carried out, is there any inviolable rule of thumb about which types of settlement are to be preferred. But the description and analysis which have been made can be carried through to two families of conclusions with practical implications: first, concerning measures to make settlement schemes more effective and economic; and second, to outline the implications in benefits, resource use, commitment and risk of settlement schemes in general, and in individual holding, compulsory marketing, scheduled production, and communal economy schemes in particular, in the hope that these may be recognized and weighed, as they rarely are, when policy choices are being made.

CHAPTER 12

Some Practical Implications

Up to this point the main concern has been to describe, compare and analyse settlement schemes. This will be justified only if it leads to better understanding and practice. Some of the implications have already been outlined and will not be repeated. Others, while not necessarily explicit, are so obvious as not to need elaboration. For instance, the transformation model used in Chapter 10 might make it easier to anticipate needs and problems in settlement schemes and other projects, particularly where roles conflict or overlap and where handovers have to take place. Again, the typology in Chapter 11 may help to show more clearly what is being chosen in deciding between different approaches to settlement, or indeed other sorts of agricultural development projects. Moreover, these frameworks may possibly provide hooks on which comparative experience of agricultural development projects could be hung intelligibly and usefully in compiling some sort of practical handbook. However not much is claimed for the transformation model and the typology. They are crude. But if they do no more than provoke others into proposing other frameworks which fit the facts more carefully, they may be minimally justified. As for this study as a whole, it may be of some benefit if, even where it is misguided or plain wrong, it stimulates some of those officials in settlement or agricultural development situations to look not just at the people among whom they work but at themselves, to try to understand their own motivations and actions, to see more clearly what they are doing and why, and to consider what they ought to do. Self-awareness and introspection of this sort can incapacitate if carried to extremes; but on a limited scale they can lead to more effective action.

Other practical policy implications must be seen in perspective against the evolution of doctrine about agricultural development schemes in general and settlement schemes in particular. In no sense at all do the lessons which emerge from the foregoing analysis displace those very important and valid lessons which have been derived earlier. It cannot be emphasized too strongly that if the organizational implications were taken on their own they would present an unbalanced set of prescriptions, focusing as they do on administrative aspects and deliberately neglecting others.

The ideas which gained currency in the 1950s included many sound insights about agricultural development projects.[1] The grandiosity and naive optimism of some of the post-Second World War schemes in colonial territories came to be recognized. It began to be realized that the scale of Gezira had been misleading, and that it could not easily be reproduced elsewhere since nowhere else in Africa were physical conditions as favourable. Mechanical cultivation in tropical conditions and underdeveloped countries was seen to suffer from major drawbacks, not least problems of implement design, spares supply and maintenance, and the creation of new labour peaks in those activities in the agricultural cycle which were not mechanized. Speed and impatience were widely blamed for the failure of schemes, and the need for extensive surveys, detailed experiments and careful pilot projects as part of gradual phased development was more fully recognized. Arthur Lewis in particular drew attention to several important factors, including selecting the right settlers, the manner in which they were inducted into schemes, their capital, the size of farming units and the tenure system.[2] At the same time settler behaviour, however unpredictable it might have appeared to expatriate managerial staff, was appreciated to be far more rational than had been supposed. The view that Africans would not know the value of money and should not be trusted with too much of it gave way to a realization that cash benefits were a prime and proper preoccupation of settlers. In accord with these ideas, the approaches to agricultural development in anglophone Africa in the latter 1950s were more piecemeal and effective and concentrated more on small schemes, extension work and the promotion of cash crops among existing smallholders than on the larger, more closely integrated and controlled projects such as settlement schemes which had been fashionable earlier.

The new surge of settlement schemes which came with independence made painfully necessary the relearning of these lessons.

---

[1] As in Alan Wood, op. cit., *passim*; Arthur Lewis, 'Thoughts on Land Settlement', *Journal of Agricultural Economics*, Vol. 11, pp. 3–11, June 1954, reprinted in Eicher and Witt, eds., op. cit., pp. 299–310; Baldwin, op. cit., esp. pp. xi–xii (Foreword by Professor Herbert Frankel) and 183–197; Gaitskell, op. cit., esp. pp. 355–357; A. H. Hanson, *Public Enterprise and Economic Development*, Routledge and Kegan Paul, 1959, *passim*. Some of these lessons are also summarized in René Dumont, *False Start in Africa*, Andre Deutsch, 1966 (original edition, *L'Afrique Noire est Mal Partie*, Editions du Seuil, Paris, 1962), pp. 52–53 and 56–59.

[2] Lewis, op. cit.

Over-ambitious plans, uneconomic mechanization, headlong implementation, and ignorance and neglect of social factors in settlement,
were almost as serious in the first half of the 1960s as they had been
in the post-war period. They remain dangers and need to be guarded
against now as much as ever. In particular, careful and appropriate
surveys and pilot projects are vital if major uneconomic investments
are to be avoided. It must be added that surveys are frequently
physical—topographical, soil, hydrological, biological, agronomic—
and rarely include market assessments or social surveys. There is a
strong case, however, for both: without a market a scheme either
dies or requires costly protection; without experience of the way
settlers or farmers will behave a project or programme cannot be
planned realistically. What is usually needed is gradual experiment
and gathering of information, and small pilot projects which include
the people to be affected. The discipline and limitations of all this
conventional wisdom are as vital now as ever, and must be accepted
by politicians and planners alike. If they are not accepted and acted
upon further extravagant mistakes will be made, mistakes which in
the long run may be disastrous to the politicians and planners
themselves.

Other considerations arising more directly from the description
and analysis of settlement schemes as organizations, will now be
outlined. They fall into two sections: first, organizational policies for
settlement schemes in which the question is, assuming there are to be
settlement schemes, how best to organize them; second, evaluating
settlement schemes, in which the question is how settlement schemes
have been and should be assessed.

Organizational policies

Some of the more conspicuous lessons for settlement scheme policies
concern departmentalism, staffing, government in business and
socio-economic measures. These will be considered in turn.

Departmentalism

Interdepartmental difficulties, although unusually acute in the early
stages of Mwea, can be regarded as a natural condition to be
expected in the formation of any institution which requires activities
contributed by different individuals or organizations. On settlement
schemes several influences in particular have given rise to friction:
fuzzy role definition, with consequent uncertainty and rivalry;
problems of co-ordination and negotiation, for instance between
designers, constructors and users; competition for resources resulting

in overlap bleeding;[1] pressures for speed leading to telescoping and bottlenecks; and problems of 'ownership' and acceptance at hand-over points. In view of all these sources of friction it is scarcely necessary to introduce personality as a major factor in explaining the difficulties experienced on Mwea, or indeed Volta. Rather such stress must be expected and anticipated on other, similar schemes. Problems as complicated and interlocking as these do not lend themselves to any simple or comprehensive solutions, but three measures can be suggested as likely to reduce them.

In the first place, all the activities required may be allocated to one man or to one department. On a small scheme it is possible for a versatile manager to carry out or supervise all the activities, as has been largely the case on Nyakashaka and Upper Kitete. Where considerable power is given to one man, there are, of course, dangers of what can be called the island syndrome: excessive possessiveness, intolerance of the outside world, resentment of 'interference' with settlers, and misanthropy. But providing the individual has the technical and administrative capacity to perform the activities required, there are substantial pay-offs: co-ordination takes place continually in one brain without committee meetings; the resource demands of the various activities are continually being assessed, and allocations are likely to be made more in the interests of a scheme as a whole than if they depended on a balance of power between different departments. Another similar way of reducing departmental difficulties is to eliminate handovers by directing or training staff to perform more than one function. For example agriculturalists, who are the ultimate managers, can perform the administrative activities of the early stages of a scheme, so that no handover is required from an organization or individual carrying out local negotiation and settler selection to an organization or individual managing a scheme. Part of the comparative success of Nyakashaka and Upper Kitete can be attributed to the fact that the managers had a large say in selecting settlers. On larger schemes such a fusion of roles is more difficult to achieve, but in evacuee and refugee schemes, for instance, it would assist the rapid achievement of settler self-sufficiency if their resettlement was carried out by agricultural staff trained in community development rather than by community development or administrative staff whose interest is more in social than economic welfare. However, on the larger and more complex schemes one man cannot always perform every specialist activity. Where this is so there is a case for making one organization responsible for all activities. If this is done specialization within the organization is inevitable, and the equivalent of interdepartmental friction

[1] See pp. 208–209.

can be expected to arise. But the presence of a co-ordinating authority should reduce stress below the level likely when activities are carried out, as they were on Mwea, by semi-autonomous departmental hierarchies, each with strong supports and allegiances outside the scheme.

Second, where specialization occurs or where there are externally-based departments creating or servicing a scheme, interaction between the staff involved should be planned, both on an organizational and on a social level. Organizationally, the lesson of Mwea does not need emphasizing. In the early stages, the lack of co-ordination was aggravated by the absence of formal consultative machinery which might have brought together departmental representatives. By the time the Local Committee was meeting regularly, departmental attitudes had hardened, and meetings were more an arena for the presentation of set departmental positions than a forum for constructive negotiation. But however important such formal meetings may be, and even if they are started in time, informal contacts and exchanges are sometimes more vital. Planning can provide for a high level of opportunity for informal contacts between the senior staff of different departments, or staff with different interests within a department, particularly through the residence patterns of staff housing and the location of offices and other buildings. Regardless of the social or national origins of staff it may be a high yielding investment for a parent scheme organization on a large scheme to provide staff with a bar, a tennis court, or a swimming pool, or some other neutral social ground which will bring staff members together.

Third, roles and handover points should be defined. There are dangers of overdefinition and inflexibility, but more problems on settlement schemes have arisen from the overlapping of the blurred edges of roles and uncertainty about handover procedures than from excessive working to role or lack of adaptability. With handovers the criteria which govern acceptance by a recipient department should be clear, and where possible abandonment of control by a withdrawing department should be made easier through formal recognition of its work. Where a withdrawing department retains responsibility for maintenance, the specification of its role is important if friction with the managing organization is to be avoided.

Staffing

Individual personality among senior staff is so crucial that it can turn an unpromising scheme into a success or a promising scheme into a failure. But the personal qualities required in a good manager are not often found. It is by no means enough that he should be a competent agriculturalist, important though that may be; he must also be a capable handler of men, and emotionally stable. It is far more difficult and frustrating to run a settlement scheme, where settlers must be persuaded and exhorted and where sanctions may be difficult to apply or self-defeating, than to operate an estate or plantation with paid labourers who may easily be dismissed. Moreover, on island schemes in particular, a manager may have to resist the temptations of power and even, as was intended with the Tanzanian Pilot Village Settlements, carry out a phased abandonment of that power by handing it over to the settlers. The psychological as well as technical demands made on settlement scheme managers are often heavy, and an adequate manager is an exceptional man.

Recruiting senior staff is therefore critical. In practice, however, attracting suitable candidates is difficult and recruiting has consequently tended to be permissive and sometimes even indiscriminate. Since settlement schemes are commonly remote and have few if any social amenities of the sort a managerial staff might expect in other occupations, candidates are often, though by no means always, those who cannot obtain better jobs or who enjoy isolation or, more exceptionally, who are inspired by idealism. A further problem is the tendency for discontinuity among senior staff, so that two types of situation are common: in which a manager remains on a scheme for some time, identifies with it, and is in danger of exhibiting the island syndrome; or in which adequate continuity for effective management is not achieved.

To reduce these problems four measures are suggested. In the first place great care is needed in recruitment. The exceptional men who make good managers can only be found either by offering exceptional inducements or by discovering men with unusual motivation. Beyond this, the sort of person who is appropriate depends on the type of scheme and its stage of development. In an individual holding scheme settlers have to be encouraged to establish themselves and major innovation is not introduced: a patient, fatherly figure is needed. In compulsory marketing schemes some of the same applies, but in addition more positive advice and control are called for: a good manager will therefore have to be able to lead, persuade and train the settlers. In scheduled production schemes the system demands the regulated and correct performance of tasks by the settlers to fit in with a centrally determined programme: a well-organized dynamic and disciplinary man is required. In communal economy schemes, in addition, there is a need to establish and sustain sufficient solidarity among settlers for the sharing of labour and rewards to be accepted and effective: an adequate manager must be technically competent, have strong humanitarian or ideological motivation, and be little short of a saint. Indeed to start a communal economy scheme without an assured succession of competent near-saints to run it is asking for trouble. Similarly, the different stages in the growth of a scheme demand different styles of management. In the earlier stages, when many activities are taking place simultaneously, a more versatile and energetic man is required than in the later stages. If and when extensive devolution to settlers takes place it may prove necessary, despite the advantages of continuity, to post in a less dynamic manager who will find it less difficult to hand over his responsibilities.

Second, since continuity among senior staff has almost invariably been less than anticipated, lack of continuity should be planned for. The harmful effects of a rapid turnover in managerial staff can be combatted in three ways: by maintaining high continuity among junior staff, as on Mwea; by organizing scheme processes as routines which can be supervised easily and which can be readily understood and accepted by junior staff and settlers, again as on Mwea; and by early devolution of responsibility to junior staff and to settlers. A combination of these three measures should improve the chances of a scheme's economic system surviving despite frequent changes among senior staff.

Third, many of the errors of over-enthusiasm, over-commitment, lack of detachment and lack of self-control which afflict settlement managers can be reduced by regular leave, and certainly by a visit to an urban centre for a night or two at least once a month. While this observation may appear dated and particularly directed towards European expatriates, there is no reason to suppose that it is not important for almost any manager of a remote scheme.

Fourth, in the colonial period the training of settlement managers was unheard of. It might, however, have reduced the failure rate of schemes considerably. In independence it is no less desirable, to pass on to managers techniques of administration, and also to help them to anticipate problems and to adapt themselves to the challenges of the different stages of development, especially the psychological obstacles to the handover of power in devolution. The meetings of Managers of Pilot Village Settlements in Tanzania, of Settlement Officers of the Kenya Million-Acre Scheme, and of Group Farm Managers in Uganda, although not necessarily designed for training as such, have had as one of their values an informal exchange of ideas and the possibility for each person to realize that his problems were neither unique nor as overwhelming as they may have appeared.

Government in business[1]

Many settlement schemes have been initated by government depart-

[1] For problems of public enterprise in less developed countries see A. H. Hanson, *Public Enterprise and Economic Development*, Routledge and Kegan Paul, 1965.

ments with government finance and staff, and have been subject to government accounting systems, audit and regulations. The example of Mwea has shown how acute the conflicts between government regulations and economic requirements can be in the case of a complex controlled irrigation scheme. It may often be most convenient for simpler projects, notably individual holding and compulsory marketing schemes, to be implemented by a government department; but with scheduled production and communal economy schemes there are many conflicts between official rules and the demands of commercial efficiency. Regulations covering stores accounting, the issue of contracts, cash accounting, and staff terms of service in government are designed to maximize accountability whereas the demands of a commercial undertaking are for flexibility to maximize efficiency, especially in production. Again, civil servants' tendencies to prefer playing safe to taking risks, observing rules to innovating, and spending votes to their limit without over-spending rather than maximizing profits, all reduce their effectiveness as managers for scheduled production or communal economy schemes. A further diseconomy is the provision of direct or indirect subsidies to the great majority of government projects. While subsidies may be justified on political or other grounds they encourage dependent attitudes on the part of staff and settlers, enable non-viable schemes to survive and often involve governments in heavy expenditure.

There is no one solution to all these difficulties. The optimum form of organization depends on a balance of factors, particularly political and administrative style and capacity. The experience of Mwea and of the National Irrigation Board in Kenya suggests, however, that if a semi-independent corporation or board can be loosely linked to government without becoming a political pork-barrel, these problems can be effectively tackled. A board or corporation taking a commercial view of settlement schemes for which it is responsible can be expected to adopt systems of accounting, staff policies, and financial measures which, by being directed towards economic viability, may be in the long-term interests not only of government but also of settlers.

Socio-economic policies

Socio-economic policies on settlement schemes are bound up with four intertwined threads: economic organization; settler security; the provision of welfare services; and the returns received by settlers for their produce. These will be considered in turn.

To take the first thread, economic organization, it has been argued above[1] that in the long run the organization of a complex settlement

<hr />

[1] See particularly pp. 164–168, 185–187, 194, 200–201 and 227–230.

scheme is determined more by its physical form and economic system than by the initial characteristics of staff or settlers. The organization and operation of scheduled production and communal economy schemes in particular come to be largely independent of the ideals or attitudes of planners, managers or settlers. Moreover, the form of organization and operation generates similar attitudes among managers and managed regardless of their ideology or origin. It makes little difference to a settler's feelings about, say, compulsory loan repayments whether they are carried out by deductions made by a distinct management or by 'his own' co-operative. The cost is the same. Nor will his responses to communal labour be much different whether it is required by a manager who has been appointed by an external agency or by a manager whom he has elected. The inconvenience and effort are the same. This is not to argue that co-operative organization is wrong, but simply that if it works it is not much different in its effects and style from a management originating outside a scheme.

The second thread is settler security. One common effect of the controls applied in scheduled production and communal economy schemes is a lack of sense of security among settlers. The greater the controls over settler activities, or the more far-reaching the sanctions, or the less the assurance of the inheritance of property—and these three factors often correlate—the less a settler is likely to feel settled. Insecurity, quite apart from its social implications, can be economically harmful, discouraging full participation in a scheme and commitment to long-term investment activities and making it less likely that a settler will work with the energy which is often associated with sense of ownership of land. The converse is also the case. Where, as in individual holding schemes, controls are slight, sanctions limited, eviction non-existent or rare, and inheritance of property assured, settlers are likely to feel more secure and to be more prepared to carry out improvements.

The third thread concerns welfare. Where high standards of housing and services are provided for settlers, whether partly as compensation (Volta), partly in order to introduce modernity (Tanzanian Pilot Village Settlements), or partly to attract settlers (Western Nigerian Settlements) they have unintended, harmful effects. In contrast, wherever low standards of services are provided and settlers have to build their own houses, clear their own land, and build their own schools, they usually show a high degree of self-reliance. At the same time, although they may consider they are neglected by government, they acquire strong feelings of ownership of the land on which they settle and a relatively high sense of security.

The fourth thread concerns settler returns for production. Settler participation and satisfaction are closely related to the cash returns they receive and the sacrifices they have to make to obtain them. Further, it is important to them that the returns should be predictable. In individual holding schemes, returns are comparatively intelligible to settlers, but less so in more complex schemes in which substantial deductions are made before payout, whether for loan repayments services, welfare or fines. If returns are unpredictable, a higher level of return may be required to maintain settler participation.

The conclusions that are drawn from these four threads or tendencies depend partly upon the view of human nature and the basis of judgment, whether social or economic, adopted by the observer. They also depend on the particular situation. In a developing country, however, there seems no justification for welfare settlements. Whatever their nominal ideology they are paternalist in spirit, since to provide settlers compulsorily with extensive services at their own cost is to deprive them of freedom to decide how to spend their money. Moreover a system in which managerial controls are limited largely to economic essentials, as a result of which settlers acquire the economic power to pay for services, is likely to be more stable and sounder than a top heavy premature welfare state in miniature in which settlers see little connection between their efforts and their rewards. In this context, Mwea is important, since it shows that high managerial control does not necessarily imply high welfare. The concentration of management from 1960 onwards on production and its concurrent resolute withdrawal from welfare functions led to a period during which there was little evidence of improvements in services to tenants or in their standards of living. But after this sag in welfare the tenants, largely of their own initiative, came together to build a self-help secondary school, and later arranged to provide part of the finance for a 60-bed hospital. Had this school and this hospital been provided for them by obligatory deductions from their payouts they would have been less satisfied, and more dependent and less active in their attitudes.

In sum, it seems reasonable to suggest that social and economic benefits from settlement schemes can be maximized more through raising production and returns than through providing services and welfare; that managerial controls should generally be limited to those that are necessary for economic efficiency; and that settler activity can be most effectively mobilized through permitting a sense of property, security, and autonomy, and through a clearly intelligible relationship between production and returns.

In the practice of management many measures, when weighed against these criteria, can be seen to have some positive and some negative effects. A manager who seeks to raise returns to settlers may find in the short run that settlers do not spend their money on social services and welfare lags behind. A manager who exercises disciplinary controls in order to increase production at the same time diminishes settlers' sense of security. A manager who seeks to achieve high returns for settlers may find that this can only be achieved at the cost of economic viability for the scheme as a whole, and conversely that economic viability can only be achieved by reducing payments to settlers. The art of management is so to balance the measures taken that the sum of their effects is to achieve all the aims—economic viability, security, welfare and high and intelligible returns. But the achievement of all these aims often proves impossible. The incompatibilities are inbuilt; they are part of the unavoidable nature of complex settlement schemes, in which solving one problem creates another and in so doing constitutes a predictable cost which is among the disadvantages of such schemes as approaches to agricultural development.

Evaluating settlement schemes

As we saw in Chapter 1 there are extensive proposals for new settlement schemes in some of the countries which fall within the scope of this inquiry. The policy lessons which have been outlined assume that settlement schemes are to be implemented; but a more fundamental and crucial question is whether there should be settlement schemes at all, and if so, of what type. The balance of political, social and economic considerations is special to each situation, but it would be wrong to shelter behind this uniqueness as an excuse for not risking generalization. In this final section, therefore, an attempt will be made to assess settlement schemes, paying particular attention to organizational aspects. These assessments will be related in turn to three approaches; first, conventional evaluation; second, the absorption of resources for which there are alternative uses; and third, the risks and commitment involved.

Conventional evaluations

There are as many criteria for evaluating settlement schemes as there are types of goals they may be intended to achieve. The three most commonly and conventionally applied are effectiveness of settlement, the achievement of political aims, and economic development. These three will be considered in turn.

From the point of view of settling people the most effective projects have been individual holding schemes. Whether populations have been displaced by dam waters, by political disturbance, by over-population or by disease, or whether potential settlers consist of migrant labourers or urban unemployed, resettlement involving a

250

high degree of self-help and limited official assistance has generally succeeded in linking people and land in relationships which are stable and which do not give rise to further problems. Compulsory marketing schemes have been somewhat less effective. The Zande Scheme disturbed the population rather than settled it. The Kenya Million-Acre Scheme has increased the population living off the land that was settled,[1] and established many smallholders, but a degree of insecurity of settlement is implied in the retention of title deeds by the Department of Settlement as security for land purchase and development loans, and the possibility this implies of eviction in case of default on repayments. With scheduled production and communal economy schemes limitations on the effectiveness of settlement have been greater. In part this has resulted from the typical failure of such schemes to achieve the targets set for them. Against the 10,000 families which in 1956 it was hoped ultimately to settle on Mwea, only 1,588 had become tenants by the end of 1966.[2] Against the target of over sixty Pilot Village Settlements each with about 250 farmers set in Tanzania in 1964 for achievement by 1969,[3] only eight, most of them only partially settled, had been created by April 1966 when the programme's suspension was announced. More directly, this failure has resulted from regimentation and control and use of sanctions, especially eviction, which are found in most scheduled production schemes, and the associated difficulty experienced by settlers in developing a sense of land ownership. The rates of turnover of settlers on different types of scheme are significant. On Chesa, an individual holding scheme, only 4 per cent (24 out of 618) had dropped out in the seven years (1957–1964) since the first arrival.[4] In contrast Orin-Ekiti, which as one of the Western Nigerian Settlements is a scheduled production scheme, the total intake for 1960–1966 inclusive was 142 settlers, but the settler population at the end of 1966 was only 60. Stability of settlement depends upon many factors, but in general seems highest on individual holding schemes and lowest on scheduled production schemes which depend on mechanization without irrigation.

A somewhat similar pattern can be seen in relation to the political aims of settlement schemes. Politically, simpler (individual holding and compulsory marketing) schemes appear generally preferable to more complex (scheduled production and communal economy) schemes. In so far as displaced populations present political problems,

---

[1] de Wilde *et al.*, op. cit., Vol. 2, pp. 218–219.
[2] M.I.S. Annual Report, 1966, p. 5.
[3] *Tanganyika Five Year Plan for Economic and Social Development, 14 July 1964–30 June 1969, Vol. 1,* Government Printer, Dar es Salaam, 1964, p. 21.
[4] Woods, op. cit., pp. 1 and 8.

the schemes which settle them effectively remove those problems and in practice these are usually individual holding schemes. In contrast, complex schemes create rather than solve political problems. Where the distribution of schemes is a form of political patronage, inducement or reward, a demand for more schemes may be aroused, particularly for those areas or constituencies which have not yet benefited. Potential settlers may ask why others and not they have been privileged by admission as settlers. The very existence of a tractor pool, or a model town, or an irrigation system, may stimulate rising demands among those who do not benefit, and when these demands cannot be met dissatisfaction with the government may result. In Tanzania the formation of a specially privileged rural settler elite conflicted with the national ethos and probably weakened its credibility among the population. The dependent attitudes that often develop among the settlers may also create continuing problems for government. Three years after the main Volta resettlement operation was completed the Ghana Government was still struggling to achieve economic self-sufficiency for many of the evacuees, and at the same time it was facing revived problems of securing adequate land for them. Indeed the collection of population into larger residential units which is a common feature of complex schemes encourages the articulation and aggregation of demands which may be politically embarrassing, whereas the scattering of population on individual holding or compulsory marketing schemes is politically stabilizing through dispersion.

Some post-independence settlement schemes can, however, be regarded as investments in stability and legitimacy. Governments need to be able to show that they are able to act and that they are taking steps to improve life for their people. Investments which are apparently non-economic may be necessary in new states in order to ensure the survival of a centralized polity which is a pre-condition for economic development.[1] The Kenya Million-Acre Scheme can be appreciated in this light. By visibly demonstrating that the Government was settling Africans on formerly European lands the Scheme created and spread a sense that a grievance was being tackled. But such effects are often short-lived and in the long run self-defeating. Those who do not benefit outnumber those who do. Stability and legitimacy may be gained in the short run by buying time, but in the long run the test for governments becomes whether or not they can continue to satisfy or contain demands, either through further similar

[1] See W. F. Ilchman and R. C. Bhargava in 'Balanced Thought and Economic Growth', *Economic Development and Cultural Change*, Vol. 14, No. 4, July 1966, pp. 385–399; they argue that development economists have overlooked this point.

measures or by other means. It is in so far as settlement schemes, whether launched for political or other reasons, become economically independent and contribute to a national economy that they augment this capacity. In the long view, then, much of the political success of a scheme depends upon its economic success.

Economic evaluations of settlement schemes have frequently been adverse. Few pre-independence schemes were subjected to rigorous economic scrutiny and the criteria of 'success' were loose and unspecified.[1] There was, as Belshaw has remarked of Uganda, an 'almost complete absence of "the economic attitude"'.[2] No economic assessments of individual holding or compulsory marketing schemes appear to have been made in the pre-independence period. With the exception of Gezira evaluations of more complex schemes were unfavourable. The Niger Agricultural Project,[3] the Sabi Valley Irrigation Projects,[4] and the Tenant Farming Scheme at Nachingwea[5] were all found wanting. Many other schemes such as the Gonja Development Company, two successive South Busoga Schemes and the Kenya Group Farms ceased to exist as schemes.

Since independence, projects have been subjected to more searching economic appraisals during and after implementation, and these have usually been unfavourable. In 1965 the verdicts of Clough and Brown on the Kenya Million-Acre Scheme, though qualified, were generally adverse, though de Wilde and others have since shown cautious optimism.[6] Wells has found that the Western Nigerian Settlements were a poor investment of resources compared with the alternatives.[7] Johnson's analysis of the Mungwi Settlement Scheme in Zambia has shown a poor record over the first five years.[8] One reason for the suspension of the Tanzanian Pilot Village Settlement Programme was the very high overheads in relation to production. In Kenya in 1966 at the time of the take-over of the three main irrigation schemes—Mwea, Perkerra and Tana—by the National Irrigation Board, it was estimated that they had shown an operating loss of over

---

[1] See for instance *African Land Development in Kenya 1946–62, passim*, but especially p. 14.

[2] Belshaw, op. cit., p. 15.

[3] Baldwin, op. cit., *passim*.

[4] Roder, op. cit., pp. 189–191.

[5] Lord, op. cit., p. 151.

[6] Clough, op. cit., and also R. H. Clough, 'Some Notes on a Recent Economic Survey of Land Settlement in Kenya', *The East African Economic Review* (New Series), Vol. 1, No. 3, December 1965, pp. 78–83; Leslie Brown in a series of articles in the *Kenya Weekly News*, 16 July–22 October 1965; de Wilde *et al.*, op. cit., Vol. 2, p. 207.

[7] Wells, op. cit.

[8] R. W. M. Johnson, 'The Northern Province Development Scheme', *Agricultural Economics Bulletin for Africa*, No. 5, April 1964 (pp. 42–110).

£1 million in 11 years.[1] Moderately favourable evaluations have been given of individual schemes, for instance of the Mweiga Settlement of the Kenya Million-Acre Scheme,[2] but these have to be weighed against many less successful projects.

Critics have qualified their strictures by conceding that most of the schemes considered had not reached maturity. Indeed, one of the problems about the evaluation of settlement schemes is that is it precisely when the largest number of activities are being carried out simultaneously that the most problems arise, the greatest attention is attracted, and the greatest concern is shown over economic viability. This is, however, too early for investments to have borne much fruit in the form of production and revenue, or even sometimes in the form of numbers of settlers put on the land. The alarm of the Kenya Treasury over Mwea in 1957 and 1958 can be interpreted in terms of expenditure in relation to those achievements which could be presented as statistics. A realistic and useful assessment should be based upon costs and benefits at maturity, that is, after the organization of production, rather than during the process of creation of a scheme.

With that one qualification, however, it will now be argued that economic evaluations should usually be even less favourable than they are. Part of the difficulty of this topic is the wide range of units which can be chosen for assessment, and the different spans of relevance which can be applied to the economic components of those units. Evaluations can be strung out along a line from most favourable to least favourable according to whether the unit of assessment is a capable or specially assisted settler; a successful scheme in a programme; a programme as a whole; an unsuccessful scheme in a programme; or an incapable or neglected settler. Similarly with spans of relevance, evaluations may be made favourable or unfavourable according to whether they ignore or include concealed overheads, hidden subsidies, interest repayments on capital, repayment of principal, protective tariffs, opportunity costs (especially of land, capital and staff) and risks such as weather, crop diseases, market fluctuations, and (in irrigation) cumulative salinity. Many of these are imponderable and difficult to quantify. This does not mean that economists are not aware of them or do not consider them, but it does mean that some at least of these factors tend to become lost in the small print, as it

[1] Memorandum by the Secretary/Accountant of the National Irrigation Board, 28 May 1966. The aggregate figures for the three schemes were capital £1,367,519; recurrent £1,644,211; depreciation £427,034; revenue £980,236; operating losses £1,091,009.

[2] For the Mweiga Scheme see Wilson Nguyo, 'Some Socio-economic Aspects of Land Settlement in Kenya', *Makerere Institute of Social Research Conference Papers, January 1967*, No. 428; and Leslie Brown, 'Contrasts in High Density Schemes', *Kenya Weekly News*, 23 July 1965, p. 7.

were, and are not represented in the final figures upon which judgment is based. Of these factors concealed and direct subsidies are especially important, and the tendency of the more complex settlement schemes to attract them, discussed above,[1] must be counted as one of their shortcomings. But beyond these weaknesses there are two other vital factors which are rarely if ever given due weight, but which should bear heavily both on decisions whether or not to implement settlement programmes and on subsequent evaluations. These are, first, certain opportunity costs; and second, the combination of risk and commitment.

Opportunity costs

Land, settlers and staff can have opportunity costs in the sense that were they not in a settlement scheme they might nevertheless be productive. Land that is occupied by a settlement scheme is sometimes, as with the black soils of Mwea, unsuitable for other uses. With irrigation in dry areas it is generally true that reliability of crops and the gross production of land are considerably increased compared with the non-irrigated condition. In other cases, other productive uses which do not involve settlement are clearly technically possible. Sometimes, as with the Kenya Million-Acre Settlements on the Kinangop plateau, it is held by agriculturalists that the pre-settlement uses of the land were more efficient than those which followed settlement. In the case of Upper Kitete, had there been no official settlement scheme, spontaneous settlement might have taken place and have achieved considerable production without heavy government involvement. The alternative to officially organized settlement schemes is often simpler, but none the less productive, land use.

Settlers' skills can also have opportunity costs. No case is known where settlers have been regarded seriously as a resource with other possible uses, but the presence of families on settlement schemes may have a cost in terms of alternative production foregone. For example, one effect of the purchase and settlement of plots on the Kenya Million-Acre Scheme by relatively prosperous Elgeyo farmers was a decline in production on the land they already owned. Moreover, settlers are often people with scarce skills. Mwea had locked up as tenants a considerable training investment in the form of masons, carpenters, mechanics, drivers, clerks, and persons who previously occupied positions of responsibility in the wage sector, some of whom might have been useful in the non-scheme economy particularly during the construction boom which was starting in Kenya in 1966. The frequency with which settlement schemes become homes for migrant workers suggests that similar costs may apply elsewhere.

[1] See pp. 197–198.

255

A more critical opportunity cost in the post-independence period has arisen from the way in which settlement schemes attract scarce junior and senior staff. The ratio of junior extension staff to settlers was 1:133 on Mwea in 1966,[1] 1:90 for the Kenya Million-Acre Scheme in 1965,[2] of the order of 1:50 for Perkerra in 1967,[3] and has been given as 1:25 for Urambo.[4] On Orin-Ekiti in Western Nigeria there were in 1966 no less than 2 Agricultural Assistants and 6 Field Oversears for only 60 settlers. In contrast the ratio of junior extension staff to farmers in Tanzania as a whole has been given as 1:1,700.[5] Similarly, the density of agricultural officers on settlement schemes has typically been very much higher than in other areas, at a time when the extension services, certainly in Kenya, Uganda and Tanzania, have been staffed well below establishment. It is remarkable that this aspect of settlement has not received greater attention. Costs and benefits have frequently been measured against units of land, capital or labour as the scarce resources, but often in developing countries it is not these factors but trained staff which are the crucial limitation. No case is known where benefits per agricultural officer or per agricultural assistant have been computed and considered. Aid agencies, in pursuing their own interests, appear blind or indifferent to the exceptionally high costs of attracting trained and locally active administrative or technical staff to special projects.[6] This point has also been overlooked by Tickner in his study of technical co-operation; although he discusses the problems of obtaining counterpart staff for technical co-operation projects, he does not consider the high indirect costs of withdrawing them from their current work.[7] A failure to pay sufficient attention to this aspect of resource use would appear to be a deficiency of agricultural economics in anglophone Africa, though Brown has cogently argued an economic case against the Kenya Million-Acre Scheme on these grounds.[8] But Brown, like others, computed the costs of staff only in terms of their direct financial costs to government. These, however, may be far lower than their opportunity costs in terms of alternative production

[1] Based on M.I.S. Annual Report, 1966, taking the ratio of Field Assistants to tenants.
[2] Leslie Brown, 'The Settlement Schemes, Staff and Continuity', *Kenya Weekly News*, 15 October 1965, p. 23.
[3] E. G. Giglioli, 'Aspects of Planning Irrigated Settlement in Kenya' (mimeo), National Irrigation Board, Nairobi, 1967, p. 9.
[4] Ruthenburg, *Agricultural Development in Tanganyika*, op. cit., p. 85.
[5] Guy Hunter, 'Development Administration in East Africa', *Journal of Administration Overseas*, Vol. 6, No. 1, January 1967, p. 12.
[6] This point is made forcefully by Hunter, ibid., *passim*, especially p. 12.
[7] Tickner, op. cit., pp. 21, 105 and 175–195.
[8] Leslie Brown, 'The Settlement Schemes, Staff and Continuity', *Kenya Weekly News*, 15 October 1965, p. 23.

foregone through not deploying them in other ways, or production lost through the disruptions caused by their transfers. These truer opportunity costs are, however, speculative and therefore difficult to quantify; and being difficult to quantify, tend to be ignored.

Of course, any assessment of settlement schemes in terms of resource use runs into difficulties of comparison with other uses. Each set of circumstances is special and each comparison unique. To consider the alternatives at all adequately would be to write another book. Agricultural extension services are the most obvious alternative to settlement schemes as a use for trained personnel but are notoriously difficult to evaluate. What can be said is that where trained and effective staff are short, a prime criterion for weighing alternative programmes for agricultural development should be the relative demands they make upon staff time and capacity; and that most settlement schemes, but especially complex ones, make heavy demands.

Risks and irreversibility of commitment

A further neglected aspect of evaluation, which again has far wider application than merely to settlement schemes, concerns the relationship between risks and irreversibility of commitment.

Settlement schemes, especially those that are more complex in system and costly in capital, are high risk undertakings. They share with non-settlement approaches to agricultural development the uncertainties of innovations, weather, pests and markets, and the disruptions of rapid turnovers in senior staff. In addition, however, they have to face other serious risks and difficulties which do not have to be borne in non-settlement situations. The land on which settlement takes place may be available for settlement for the simple reason that it is marginal or unsuitable for cultivation. The locations of many settlement schemes, often with poor communications and far from the services of urban centres, raise costs and the difficulties of management. There is a danger that both organizational and productive effectiveness will be restrained by the inbuilt incompatibilities of complex schemes, by the cancelling out of managerial and settler efforts, in the games of enforcing and beating the system. Moreover, adaptations of schemes to ensure the continued presence and participation of settlers may have to be made through increased payouts or through services which at best reduce revenue to government, and at worst add to a loss. In addition, where government withdrawal is intended there is a high risk that it will take longer than expected. At the point at which implementation of a settlement scheme or programme is considered all these factors, all of them implying economic risks, should be weighed.

But these risks do not present the complete picture. Wherever a government starts a programme or project the actual risks are compounded by the extent to which the commitment to maintain the programme or project is irreversible. The process of commitment can be lengthy, subtle and insidious. It begins with an opportunity and a vision. These may arise from a disturbance in the relationships of men and land, or the perception of unoccupied land: the Mwea, inviting development after the Kenya Land Commission's recommendations; the bush of South Busoga after its evacuation in the first decade of this century; the narrow strip of uncultivated land on the edge of the Rift Valley at Upper Kitete; the cleared bush of Kongwa, Urambo and Nachingwea after the Groundnut Scheme fiasco. Or the opportunity may be provided by a resettlement operation which presents a captive population which can be directed into a new agricultural system: the displacement of Halfawis by the Aswan dam was exploited through resettlement on the controlled irrigation scheme at Khasm-el-Girba; and the evacuees from the Volta lake were thought to provide 'a unique opportunity to wean an appreciable proportion of Ghana's farmers from the wasteful, fragmented, and shifting system of agriculture to a settled and improved pattern of farming'.[1] The opportunity attracts and nourishes the idea of a scheme. In such conditions a personal commitment can develop in a man of vision like Simon Alvord in Rhodesia or Chief Akin Deko in Nigeria. Funds are obtained for surveys: the surveys that are carried out are themselves committing. Where their findings are marginal, as was the United Nations Special Fund survey of the proposed Tana Irrigation Project in Kenya,[2] further investigations are called for. It becomes increasingly difficult to turn back. Once funds have been made available for a substantive scheme, the successive activities of planning, construction, settlement and production draw after them deeper and deeper personal, departmental and political commitments. The establishment of settlers sets a seal on commitment at a higher level, making abandonment extremely difficult and the use of protective political arguments extremely easy. The full repertoire of defences to ensure scheme or programme survival can now be brought into play. Moreover officials and politicians in circumstances such as these may regard government funds as fair game, as an ecological feature to be exploited much as a river might be tapped for irrigation water. The risks involved in the original initiation of a project are now more obvious: risks not merely that it

[1] Nicholas, in *Volta Resettlement Symposium Papers*, op. cit., p. 86.
[2] Interim report of a United Nations Special Fund survey of the lower Tana basin in Kenya, as reported in *East African Standard*, (Nairobi), 17 November 1965, and *Daily Nation*, (Nairobi) 17 November 1965.

would fail, but that having failed it would survive as a parasite that could not be shaken off or killed.

An example is the Perkerra Irrigation Scheme in Kenya, a project which in the past has been uneconomic by almost any standards but which has not been folded up, despite heavy operating losses. When the issue of closing down was raised in 1959, it was decided to run the scheme on a care and maintenance basis for three years and decide its future at the end of that period.[1] But by 1962 closure had become more difficult and the scheme was expanded in an attempt to make it less of a recurrent burden on government, thus deepening commitment. In 1966 the National Irrigation Board was then saddled with meeting an operating loss of the order of £20,000 per annum in order to keep the Scheme in existence and in 1967, with further expansion, the number of tenants passed the 500 mark, making closure even more difficult to contemplate. Moreover, the combination of able management, a heavily protected domestic onion market, intermittent good luck, and the most favourable possible methods of economic evaluation disguise the fact that it is a drain on resources which might perhaps be used in other ways more productively for a larger number of people.

The issues involved in a decision to terminate a scheme are, of course, not simple. Those responsible for the decisions may not even agree about whether the amount of money already sunk in a project is a relevant consideration. Attitudes and ideas are sufficiently confused and contradictory for irrational elements to have free play. It is extremely difficult, for example, to see the large quantities of fine onions grown on Perkerra, and to compare the green irrigated fields with the surrounding desert, and at the same time to sustain a conviction that the Scheme should be abandoned. Running water through channels and on to dry land, growing abundant crops where there was only bare soil and barren bush before, and enabling people to enjoy a level of prosperity they have never previously known, appear inherently and incontrovertibly good. It is Isaiah's vision:

> The wilderness and the solitary place shall be glad for them; and the desert shall rejoice, and blossom as the rose.[2]

To suggest closure seems ignoble and destructive, an affront to the aspirations and achievements of the human spirit. If a neutral visitor can have this feeling, it may be expected all the more in those whose lives and work are bound up in a scheme. Yet the power of this emotion multiplies the risks of starting projects of this sort through

---

[1] *Department of Agriculture Annual Report 1959*, p. 51.
[2] Isaiah, chapter 35, verse 1.

making it exceptionally difficult to close them down however uneconomic they may prove.

There is, indeed, a strain of Utopianism in most complex settlement schemes. Often there is an idealized view of the human situation that settlement will create. In colonial times this was often the stabilized African, fixed and controlled on a piece of land. Since independence, it has varied: in West Africa it has been an urban farmer; in Kenya, a sturdy yeoman; in Tanzania, a co-operative worker. Another Utopian aspect is the frequency with which stresses and breakdowns are not anticipated: as Apthorpe has pointed out, provision is often lacking either for failure of the social system or for mechanical repairs.[1] Again, it is very common for the targets for land preparation, settlement, production and withdrawal to be wildly optimistic and for achievements to fall far short of them. These features are partly explicable in terms of the self-delusion of men who are transported by a vision. When an ideal is pursued by a whole community, as in some communal economy schemes, it may make a scheme feasible through the sacrifices the participants are prepared to accept; but when the vision is only in the mind of the initiator, as it has been with most complex settlement schemes, the effects are often a sequence of unrealistic estimates, uneconomic measures and personal commitments which comprise part of the risks of the project.

Concluding

Since all these disadvantages have applied in the past they can be expected to continue to apply in the future, and should be taken into account in assessing proposals for settlement schemes and similar agricultural projects. It is not enough to carry out evaluations which consider only those economic factors which can be quantified; it is necessary also to include administrative factors and the probable motivations and behaviour of the actors involved. Allowance has to be made, for example, for the expected patterns of settler and managerial behaviour, for departmentalism, for staff discontinuities and for the inbuilt incompatibilities of scheme systems. Moreover, comparisons with alternative approaches to agricultural development should take into account the high opportunity costs of trained staff and the expected ease or difficulty of abandoning a project or programme if it proves uneconomic. When all this is done the case against high capital and complex settlement schemes becomes stronger than when only conventional cost-benefit criteria are used. While this does not mean that such schemes should be ruled out

[1] Apthorpe, *East African Institute of Social Research Conference Papers, January 1966*, op. cit., p. 23.

altogether, it does mean that they should be approached with greater care and understanding.

Where a settlement scheme is unavoidable, and where there is a choice of type to be adopted, there is much to be said on organizational grounds for the simplest type of scheme that is compatible with the circumstances of settlement. The simpler approaches are relatively undemanding of scarce administrative and technical capacity, and engage it for shorter periods. They involve relatively low risk and low commitment. Moreover, schemes with individual holdings exploit the drives of property ownership and individual incentive which can make productive the labour which is the most abundant unused resource in much of the third world. The simpler schemes also require intermediate levels of organization corresponding with the intermediate technology which may also be appropriate. If the beginning is ambitious, a complex organization may collapse and find equilibrium at a lower level; but if the beginning is modest, a more complex technology and organization can grow up organically and gradually. For example, the tractors appearing on Chesa in Rhodesia and on the Kenya Million-Acre Schemes as a result of settler initiative represent a self-sustaining upward movement in which productivity may increase without heavy government investment or commitment. If such developments are to be possible, it is important that advisory and technical services be available when needed, and even more important that the system of land tenure adopted should allow for future flexibility in farm size. Given such flexibility, it is usually safer and sounder to develop piecemeal from an existing base, whether this is farmers already on their land or settlers already on a scheme, than to attempt radical transformation in one long step.

Settlement schemes, particularly those which are complex in system, will remain temptations. Because of their creative possibilities, they will continue to find energetic and enthusiastic sponsors. Because of their visibility, clear boundaries, organizational coherence, and Utopian overtones, they will no doubt continue to attract successive colonization—by administrators who negotiate their emergence, constructors who build them, agriculturalists who manage them, settlers who populate them, and in their wake foreign aid personnel and research students[1] in various capacities—all of whom will find satisfaction in occupying a bounded and identifiable territory. What is vital is not that such schemes should be avoided on principle, but that those who act in these situations should appreciate what is happening. It is especially important that those who make development decisions should understand themselves well enough to be able to compensate in their acts of judgment for the

[1] *Mea culpa.*

261

strong pull of the psychological attractions of such schemes, and should be able to see clearly the risks they entail and the benefits that might accrue from alternative uses of the resources involved. Exceptional restraint and imagination are needed among politicians and civil servants if the lure of the big scheme is to be neutralized so that a balanced and realistic assessment can be made. Perhaps it is fortunate that so many African politicians and civil servants possess and farm their own land. While this may be distraction it may also satisfy desires for property and territory, so that they are less prone than their expatriate predecessors to seek such satisfaction through their work. It may in the long term enable them to take more balanced views of policy and to appreciate more fully the alternatives that exist. Certainly it is important to recognize that the choices are neither clearcut nor easy. It is not enough, as was done in Kenya before independence,[1] to quote Gulliver's report of the views of the King of Brobdignag:

> And he gave it for his opinion, that whoever could make two ears of corn or two blades of grass to grow upon a spot of ground where only one grew before, would deserve better of mankind, and do more essential service to his country than the whole race of politicians put together.[2]

For the issues are less simple: they include whether, with the same resources, many more ears of corn, or many more blades of grass, might not be grown in other ways or in other places; and whether those politicians and civil servants who make major policy decisions have the freedom, the insight and the courage to choose those other ways or places, however unspectacular they may be.

[1] *African Land Development in Kenya 1946–1962*, op. cit., p. 1.
[2] Jonathan Swift, *Gulliver's Travels* (first published 1726), Chapter 7.

Bibliography

Each section is divided into (a) Books, (b) Monographs and Articles, (c) Official Publications, (d) Unpublished Papers.

Under A (a) Books, general: are listed books of a general nature, books which refer to settlement schemes in more than one African country, and books which, while about a settlement scheme in a particular country, have a notable wider relevance.

Under the country headings are listed books which are relevant only to those countries, and sections of some other books which deal with settlement schemes in those countries. Where the latter have already been listed under A (a), a short reference only is given.

CONTENTS

A. GENERAL

a. *Books*

ADU, A. L., *The Civil Service in New African States*, Allen & Unwin, 1965

ALLAN, W., *The African Husbandman*, Oliver and Boyd, Edinburgh, 1965

ARDREY, ROBERT, *The Territorial Imperative, a Personal Inquiry into the Animal Origins of Property and Nations*, Collins, 1967

BALDWIN, K. D. S., *The Niger Agricultural Project, an Experiment in African Development*, Blackwell, 1957

BASCOM, WILLIAM R., and HERSKOVITS, MELVILLE J. (eds.), *Continuity and Change in African Cultures*, University of Chicago Press, 1959

BATES, W. N., *Mechanization of Tropical Crops*, Temple Press Books, London, 1957

BELOV, FEDOR, *The History of a Soviet Collective Farm*, Praeger, New York, undated

T   263

BIEBUYCK, DANIEL (ed.), *African Agrarian Systems* (studies presented and discussed at the Second International African Seminar, Lovanium University, Leopoldville, January 1960), Oxford University Press, 1963

BOHANNAN, PAUL, *African Outline, a General Introduction*, Penguin Books, 1966

BURNS, TOM and STALKER, G. M., *The Management of Innovation*, Tavistock Publications, 1961

DE GRAFT-JOHNSON, J. C., *African Experiment, Co-operative Agriculture and Banking in British West Africa*, Watts, London, 1958

DE WILDE, JOHN C. et al., *Experiences with Agricultural Development in Tropical Africa: Volume 1 The Synthesis* and *Volume 2 The Case Studies*, The John Hopkins Press for the International Bank for Reconstruction and Development, Baltimore, 1967

DUMONT, RENÉ, *False Start in Africa* (translated by Phyllis Nauts Ott), Andre Deutsch, 1966 (originally published as *L'Afrique Noire est Mal Partie*, Editions du Seuil, Paris, 1962)

EICHER, CARL K., and WITT, LAWRENCE W. (eds.), *Agriculture in Economic Development*, McGraw-Hill, New York, 1964

EISENSTADT, S. N., *The Absorption of Immigrants, a Comparative Study Based Mainly on the Jewish Community in Palestine and the State of Israel*, Routledge & Kegan Paul, 1954

ETZIONI, AMITAI (ed.), *Complex Organizations, a Sociological Reader*, Holt, Rinehart & Winston, New York, 1962

ETZIONI, AMITAI and ETZIONI, EVA, *Social Change—Sources, Patterns and Consequences*, Basic Books, New York, 1964

FARMER, B. H., *Pioneer Peasant Colonization in Ceylon, a Study in Asian Agrarian Problems*, Oxford University Press, 1957

FINKLE, JASON L., and GABLE, RICHARD W. (eds.), *Political Development and Social Change*, Wiley, New York, 1966

FIRTH, RAYMOND, and YAMEY, B. S. (eds.), *Capital, Saving and Credit in Peasant Societies*, Allen & Unwin, 1964

FOSTER, GEORGE M., *Traditional Cultures and the Impact of Technological Change*, Harper & Row, New York, 1962

GAITSKELL, ARTHUR, *Gezira, a Story of Development in the Sudan*, Faber & Faber, 1959

GRAY, ROBERT F., *The Sonjo of Tanganyika, an Anthropological Study of an Irrigation-based Society*, Oxford University Press, 1963

HAILEY, LORD, *An African Survey, a Study of Problems Arising in Africa South of the Sahara*, Oxford University Press, 1938

— *An African Survey (Revised 1956), a Study of Problems Arising in Africa South of the Sahara*, Oxford University Press, 1957

HANSON, A. H., *Public Enterprise and Economic Development*, Routledge & Kegan Paul, 2nd edition with new preface, 1965

HELLEINER, G. K. (ed.), *Agricultural Planning in East Africa*, East African Rural Development Studies I, East African Publishing House, Nairobi, 1968

HEUSSLER, ROBERT, *Yesterday's Rulers, the Making of the British Colonial Service*, Syracuse University Press, Syracuse, 1963

264

HINDEN, RITA (ed.), *Co-operation in the Colonies, a Report from a Special Committee to the Fabian Colonial Bureau*, Allen & Unwin, 1945

HUNTER, GUY, *The New Societies of Tropical Africa, a Selective Study*, Oxford University Press, 1962

— *The Best of Both Worlds? A Challenge on Development Policies in Africa*, Institute of Race Relations, Oxford University Press, 1967

HUXLEY, ELSPETH, *Forks and Hope, an African Notebook*, Chatto & Windus, 1964

JAHODA, GUSTAV, *White Man, a Study of the Attitudes of Africans to Europeans in Ghana before Independence*, Institute of Race Relations, Oxford University Press, 1961

JEFFRIES, SIR CHARLES, *The Colonial Office*, Allen & Unwin, 1956

KANOVSKY, ELIYAHU, *The Economy of the Israeli Kibbutz*, Harvard University Press, Cambridge, Mass., 1966

KIMBLE, GEORGE H. T., *Tropical Africa* (2 vols), The Twentieth Century Fund, New York, 1960

KREININ, MORDECHAI E., *Israel and Africa, a Study in Technical Co-operation*, Praeger, New York, 1964

LEE, J. M., *Colonial Development and Good Government, a Study of the Ideas Expressed by the British Official Classes in Planning Decolonization 1939–1964*, Oxford University Press, 1967

LORENZ, KONRAD, *On Aggression* (translated by Marjorie Latzke), Methuen, 1967 (first published in Vienna by Borotha-Schoeler Verlag under the title *Das Sogenannte Bose*)

LOWE-MCCONNELL, R. H., *Man-made Lakes* (Proceedings of a Symposium held at the Royal Geographical Society, London, 30 September and 1 October 1965), Academic Press, London, 1966

MAIR, LUCY, *Primitive Government*, Penguin Books, 1962

MANNONI, O., *Prospero and Caliban, the Psychology of Colonization* (translated by Pamela Powesland), Praeger, New York, 1956 (first published in 1950 by Editions du Seuil, Paris, under the title *Psychologie de la Colonisation*)

NICULESCU, BARBU, *Colonial Planning, a Comparative Study*, Allen & Unwin, 1958

REINING, CONRAD C., *The Zande Scheme, an Anthropological Case Study of Economic Development in Africa*, Northwestern University Press, Evanston, Illinois, 1966

RODER, WOLF, *The Sabi Valley Irrigation Projects*, University of Chicago Department of Geography Research Paper No. 99, University of Chicago Press, 1965

SELZNICK, PHILIP, *TVA and the Grass Roots*, University of California Press, Berkeley, 1949

SPENCER, PAUL, *The Samburu, a Study of Gerontocracy in a Nomadic Tribe*, Routledge & Kegan Paul, 1965

SPICER, EDWARD H. (ed.), *Human Problems in Technological Change, a Casebook*, Russell Sage Foundation, New York, 1952

SPIRO, MELFORD, E., *Kibbutz, Venture in Utopia*, Schocken Books, New York, 1956

265

SWERDLOW, IRVING (ed.), *Development Administration, Concepts and Problems*, Syracuse University Press, Syracuse, 1963

SWIFT, JONATHAN, *Gulliver's Travels*, Oxford University Press, 1935 (first published 1726)

SYMONDS, RICHARD, *The British and Their Successors, a Study in the Development of the Government Services in the New States*, Faber & Faber, 1966

TICKNER, F. J., *Technical Co-operation*, Hutchinson, 1965

WARRINER, DOREEN, *Land Reform and Development in the Middle East, a Study of Egypt, Syria and Iraq*, Royal Institute of International Affairs, London, 1957

WEINGROD, ALEX, *Reluctant Pioneers, Village Development in Israel*, Cornell University Press, Ithaca, 1966

WHETHAM, EDITH H., *Co-operation, Land Reform and Land Settlement: Report on a Survey in Kenya, Uganda, Sudan, Ghana, Nigeria and Iran*, The Plunkett Foundation for Co-operative Studies, London, 1968

WITTFOGEL, KARL A., *Oriental Despotism, a Comparative Study of Total Power*, Yale University Press, New Haven, 1957

WOOD, ALAN, *The Groundnut Affair*, The Bodley Head, 1950

WORTHINGTON, E. B., *Science in the Development of Africa, a Review of the Contribution of Physical and Biological Knowledge South of the Sahara* (prepared at the request of the Commission for Technical Co-operation South of the Sahara (C.C.T.A.) and the Scientific Council for Africa South of the Sahara (C.S.A.)), no publisher given, undated but from internal evidence probably 1958

WRIGLEY, GORDON, *Tropical Agriculture, the Development of Production*, Batsford, 1961

YUDELMAN, MONTAGUE, *Africans on the Land, Economic Problems of African Agricultural Development in Southern, Central and East Africa with Special Reference to Southern Rhodesia*, Harvard University Press, Cambridge, Mass., 1964

b. *Monographs and articles*

APTHORPE, RAYMOND, 'A Survey of Land Settlement Schemes and Rural Development in East Africa', *East African Institute of Social Research Conference Papers*, January 1966, No. 352

APTHORPE, RAYMOND, ed., 'Land Settlement and Rural Development in Eastern Africa', *Nkanga Editions*, No. 3, Transition Books, Kampala, undated

BEN-DAVID, JOSEPH (ed.), *Agricultural Planning and Village Community in Israel*, UNESCO, Paris, 1964

BRIDGER, G. A., 'Planning Land Settlement Schemes', *Agricultural Economics Bulletin for Africa*, No. 1, September 1962

CHRISTODOULOU, D., 'Land Settlement: Some Oft-neglected Basic Issues', *Monthly Bulletin of Agricultural Economic Statistics*, F.A.O., Rome, Vol. 14, No. 10, 1965, pp. 1-6

CLARK, COLIN, *The Economics of Irrigation in Dry Climates*, Institute for Research in Agricultural Economics, Oxford, undated, but post-1954

DANIEL, J. B. MCI., 'Some Government Measures to Improve African Agriculture in Swaziland', *Geographical Journal*, Vol. 132, Part 4, December 1966, pp. 506–515

DU SAUTOY, PETER, 'Resettlement Schemes and Community Development', *Community Development Bulletin*, Vol. 12, No. 4, September 1961, pp. 121–123

DUMPLETON, C. W., 'Colonial Development Corporation', *Fabian Research Series 186*, Fabian Colonial Bureau, London, 1957

HARRISON CHURCH, R. J., 'Observations on Large Scale Irrigation Development in Africa', *Agricultural Economics Bulletin for Africa*, No. 4, November 1963

HUNTER, GUY, 'Development Administration in East Africa', *Journal of Administration Overseas*, Vol. 6, No. 1, January 1967, pp. 6–12

HUSSAIN, ALTAF, 'Problems in the Planning of Land Settlement Programmes and Their Integration in the Overall Development Plan of a Country', *Proceedings of the Conference of the Philosophical Society of the Sudan*, December 1965, Vol. 2, pp. 79–97

ILCHMAN, W. F., and BHARGAVA, R. C., 'Balanced Thought and Economic Growth', *Economic Development and Cultural Change*, Vol. 14, No. 4, July 1966, pp. 385–399

JACOBS, BENET, '13 Points on the Problems of Administration in New African States', *Transition*, September/October 1964, pp. 31–32

JOLLY, A. L., 'Group Farming', *Tropical Agriculture*, Vol. 27, July/December 1950, pp. 150–153

LEWIS, W. ARTHUR, 'Thoughts on Land Settlement', *Journal of Agricultural Economics*, Vol. 11, June 1954, pp. 3–11

MORIS, JON, 'The Evaluation of Settlement Scheme Performance: A Sociological Appraisal', *Makerere Institute of Social Research Conference Papers*, January 1967, No. 430

Papers of the Conference on the Legal, Economic and Social Aspects of African Refugee Problems, (held in Addis Ababa in October 1967), United Nations High Commission for Refugees, Geneva, 1967

ROSTOW, W. W., 'Economic Development, the Importance of Agriculture', *Kenya Weekly News*, 26 March 1965

SEGAL, AARON, 'Refugees in Africa', *Kenya Weekly News*, 22 January 1965

TANNER, R. E. S., 'Conflict within Small European Communities in Tanganyika', *East African Institute of Social Research Conference Papers*, July 1962

WALLMAN, SANDRA, 'The Farmech Scheme: Basutoland (Lesotho)', *African Affairs*, Vol. 67, No. 267, April 1968, pp. 112–117

WEINGROD, ALEX, 'Administered Communities: Some Characteristics of New Immigrant Villages in Israel', *Economic Development and Cultural Change*, Vol. 11, No. 1, October 1962

WHITE, GILBERT F., 'Contributions of Geographical Analysis to River Basin Development', *Geographical Journal*, Vol. 129, Part 4, December 1963, pp. 412–436 (includes bibliography)

World Land Reform Conference Papers, 20 June–2 July 1966, F.A.O., Rome, 1966

YELD, R., 'Implications of Experience with Refugee Settlement', *East African Institute of Social Research Conference Papers*, 1965

c. *Official publications*

Report and Proceedings of the Conference of Colonial Directors of Agriculture held at the Colonial Office, July 1931, Colonial No. 67, H.M.S.O., 1931

E. HARRISON, *Soil Erosion*, Crown Agents, London, 1937

Report and Proceedings of the Conference of Colonial Directors of Agriculture held at the Colonial Office, July 1938, Colonial No. 156, H.M.S.O., London, 1938

Colonial Agricultural Policy, C. M. No. 9, Colonial Office, January 1945

The Co-operative Movement in the Colonies, Despatches dated 20 March 1946 and 23 April 1946 from the Secretary of State for the Colonies to Colonial Governments, Colonial No. 199, H.M.S.O., London, 1946

A Plan for the Mechanized Production of Groundnuts in East and Central Africa, Cmnd. 7030, H.M.S.O., London, February 1947

Agriculture in the Colonies, Colonial Office, London, July 1947

Colonial Primary Products Committee Interim Report, Colonial No. 217, H.M.S.O., London, 1948

LYNN, C. W., *Agricultural Extension and Advisory Work with Special Reference to the Colonies*, Colonial No. 241, H.M.S.O., London, 1949

Agricultural Development in Africa, Colonial Office Summer Conference on African Administration, 3rd Session, African No. 117, H.M.S.O., London, 1949

Land and Population in East Africa, Colonial No. 290, H.M.S.O., London, 1952

Report of the East African Royal Commission 1953/55, Cmd. 9475, H.M.S.O., London, 1955

Notes on Some Agricultural Development Schemes in British Colonial Territories, H.M.S.O., London, October 1955

d. *Unpublished papers*

APTHORPE, RAYMOND, 'Land Settlement and Rural Development: Part 1 Some Definitional and Conceptual Problems, Models and Phases', paper to Conference on Education, Employment and Rural Development, University College, Nairobi, 1966

'A Tentative Checklist of Questions about Settlement Schemes' (mimeo), 'based on initial discussions of Brain, Charsley, Chambers, Robertson and Yeld at Makerere on 5 January 1966 and subsequent discussions of Apthorpe, Brown, Chambers, Moris, Myers, Nellis, Rigby and Yeld, with assistance from Etherington, at Nairobi on 4 and 5 February 1966. Edited by Hutton and Apthorpe, following pre-editing by Chambers'

BETTS, T. F., (Field Director, Oxfam) 'Refugees in Eastern Africa, a Comparative Study' (mimeo), Nairobi, 6 May 1966

CHAMBERS, ROBERT, 'Phases in Settlement Schemes, Some Preliminary Thoughts', paper to Seminar on Rural Development, Syracuse University and University College, Dar es Salaam, 4 to 7 April 1966

Colonial Service Second Devonshire Course, Cambridge 1952–53, 'Land Settlement in the British West Indies' (mimeo)

Colonial Service Second Devonshire Course, Cambridge 1952–53, 'Rural and Urban Settlement in the Colonies: Report of the Study Group on West Africa' (mimeo)

MORIS, JON R., 'The Educational Requirements of a Transformational Approach to Agricultural Development', paper to Seminar on Rural Development, Syracuse University and University College, Dar es Salaam, 4 to 7 April 1966

ROSENFELD, EVA, 'Institutional Change in Israeli Collectives', unpublished Ph.D thesis, Columbia University, 1952

SEGAL, AARON, 'The Politics of Land in East Africa' (mimeo), paper to the Conference of the American African Studies Association, 1966

VALLIER, IVAN ARCHIE, 'Production Imperatives in Communal Systems: a Comparative Study with Special Reference to the Kibbutz Crisis', unpublished Ph.D thesis, Harvard, 1959

WALLMAN, SANDRA, 'Farmech Mechanization Scheme, Mafeteng District, Lesotho (1961 to date)' (mimeo), paper to a symposium of the African Studies Association of the United Kingdom, 22 September 1967

B. GHANA

a. *Books*

AMARTEIFIO, G. W., BUTCHER, D.A.P., and WHITHAM, DAVID. *Tema Manhean, a Study of Resettlement*, Planning Research Studies Number Three, published for the University of Science and Technology, Kumasi by Ghana Universities Press, Accra, 1966

ANYANE, S. LA., *Ghana Agriculture, its Economic Development from Early Times to the Middle of the Twentieth Century*, Oxford University Press, 1963

LOWE-MCCONNELL, R. H., *Man-made Lakes*, pp. 99–108, Thayer Scudder, 'Man-made Lakes and Population Resettlement in Africa', 1966

WHETHAM, EDITH H., *Co-operation, Land Reform and Land Settlement*, 1968, pp. 38–42

b. *Monographs and articles*

BROKENSHA, DAVID (ed.), '*Volta Resettlement, Ethnographic Notes on Some of the Southern Areas*' (mimeo), Legon, Ghana, Department of Sociology, University of Ghana, April 1962

BROKENSHA, DAVID, 'Volta Resettlement and Anthropological Research', *Human Organization*, Vol. 22, 1963, pp. 286–290

DE GRAFT JOHNSON, J. C., *Background to the Volta River Project*, Abura Printing Works, Kumasi, 1955

JACKSON, SIR ROBERT, 'The Volta River Project', *Progress*, Vol. 50, No. 282, 1964, pp. 146–161

LAWSON, ROWENA, 'The Structure, Migration and Resettlement of Ewe Fishing Units', *African Studies*, Vol. 17, No. 1, 1958, pp. 21–27

— 'An Interim Economic Appraisal of the Volta Resettlement Scheme', *The Nigerian Journal of Economic and Social Studies*, Vol. 10, No. 1, March 1968, pp. 95–109

— 'The Volta Resettlement Scheme', *African Affairs*, Vol. 67, No. 267, April 1968, pp. 124–129

Volta Resettlement Symposium Papers, (papers read at the Volta Resettlement Symposium held in Kumasi, 23–27 March 1965) (mimeo), Volta River Authority, Accra, and Faculty of Architecture, University of Science and Technology, Kumasi, 1965

'Waiting for Valco', *West Africa*, No. 2586, 24 December 1966, p. 1475

c. *Official publications*

The Volta River Project Vol. 1: Report of the Preparatory Commission, H.M.S.O., London, 1956

The Volta River Project Vol. 2: Appendices to the Report of the Preparatory Commission, H.M.S.O., London, 1956

The Volta River Project, Statement by the Government of Ghana, 20 February 1961, Government Printer, Accra, 1961

The Volta River Project, a Statement by Osagyefo Dr. Kwame Nkrumah, President of the Republic of Ghana, to the National Assembly, February 21 1961, Ghana Information Services, Accra

Volta River Development Act, 1961, Act 46, Government Printer, Accra, 1961

Annual Report of the Department of Social Welfare and Community Development, 1961, Government Printing Department, Accra, 1963 (also Annual Reports for 1962–1964)

Volta River Authority Annual Report for the Year 1963, Guinea Press, Accra (also Annual Reports for 1964 and 1965)

Agency for International Development, *Evaluations, Suggestions, and Recommendations on Resettlement of Volta River Area, Ghana*, Washington, D.C., 1963

Seven-Year Plan for National Reconstruction and Development 1963/64– 1969/70, Government Printing Department, Accra, 1964

b. *Unpublished Papers*

Quarterly Reports of the Volta River Authority (mimeo), December 1962 to March 1965 inclusive

LYIMO, S. N. P., and MSELE, J. D., 'Report on the Construction of Unauthorized Structures in New Ajena' (mimeo), 15 August 1964

'Revised Village Groupings' (mimeo), published by the Survey Section, Volta River Authority, 2nd revision, 27 February 1965

Materials on Social Survey (mimeo), Resettlement Office, Volta River Authority, undated

LAWSON, ROWENA M., 'An Interim Economic Appraisal of the Volta Resettlement Scheme—Ghana' (mimeo), paper to a symposium of the African Studies Association of the United Kingdom, 22 September, 1967

C. KENYA

(i) GENERAL

a. *Books*

ALLAN, WILLIAM, *The African Husbandman*, 1965, pp. 392–406

BENNETT, GEORGE, *Kenya, A Political History, the Colonial Period*, Oxford University Press, 1963

BENNETT, GEORGE, and ROSBERG, CARL, *The Kenyatta Election: Kenya 1960–1961*, Oxford University Press, 1961

CAREY JONES, N. S., *The Anatomy of Uhuru, an Essay on Kenya's Independence*, Manchester University Press, 1966

CLAYTON, ERIC, *Agrarian Development in Peasant Economies, Some Lessons from Kenya*, Pergamon Press, 1964

HUXLEY, ELSPETH, *A New Earth, an Experiment in Colonialism*, Chatto & Windus, 1960

KENYATTA, JOMO, *Facing Mount Kenya, the Tribal Life of the Gikuyu*, Secker & Warburg, 1938

LAMBERT, H. E., *Kikuyu Social and Political Institutions*, Oxford University Press, 1956

MIDDLETON, JOHN and KERSHAW, GREET, *The Kikuyu and Kamba of Kenya*, Ethnographic Survey of Africa, East Central Africa, Part V, International African Institute, London, 1965

ROSBERG, CARL G. JR., and NOTTINGHAM, JOHN, *The Myth of 'Mau Mau': Nationalism in Kenya*, Praeger, New York, 1966

RUTHENBERG, HANS, *African Agricultural Production Development Policy in Kenya 1952–1965*, Springer-Verlag, Berlin, 1966

SORRENSON, M. P. K., *Land Reform in the Kikuyu Country, a Study in Government Policy*, Oxford University Press, Nairobi, 1967

WHETHAM, EDITH H., *Co-operation, Land Reform and Land Settlement*, 1968, pp. 11–22

b. *Monographs and articles*

CHAMBERS, ROBERT, 'Some Background to Proposals for Development Plan Implementation at Provincial and Lower Levels in Kenya', *East African Institute of Social Research Conference Papers*, January 1966, No. 381

Chief Secretary of Kenya's circular No. 12 of 1960, reproduced in *Journal of African Administration*, Vol. 13, No. 1, January 1961, pp. 50–52

EVANS, M. N., 'Local Government in the African Areas of Kenya', *Journal of African Administration*, Vol. 7, No. 3, July 1955, pp. 123–127

GERTZEL, CHERRY, 'The Provincial Administration in Kenya', *Journal of Commonwealth Political Studies*, Vol. 4, No. 3, November 1966, pp. 201–215

HOMAN, F. D., 'Consolidation, Enclosure and Registration of Title in Kenya', *Journal of Local Administration Overseas*, Vol. 1, No. 1, January 1962, pp. 4–14

PENWILL, D. J., 'Paper—The Other Side', *Journal of African Administration*, Vol. 6, No. 3, July 1954, pp. 115–123

'Planned Group Farming in Nyanza Province, Kenya', article published by courtesy of the Director of Agriculture, Kenya, *Tropical Agriculture*, Vol. 27, July–December 1950, pp. 153–157

SORRENSON, H. P. K., 'The Official Mind and Kikuyu Land Tenure (1895–1939)', *East African Institute of Social Research Conference Paper*, January 1963

— 'Counter-Revolution to Mau Mau: Land Consolidation in Kikuyuland, 1952–1960' (mimeo), *East African Institute of Social Research Conference Papers*, June 1963

c. *Official publications*

Kenya Land Commission: Evidence and Memoranda, Vol. 1, Colonial No. 91, H.M.S.O., London, 1934

Report of the Kenya Land Commission, Colonial No. 4556, H.M.S.O., London, 1934

Native Lands Trust Ordinance, cap. 100, Government Printer, Nairobi, 1949

A Plan to Intensify the Development of African Agriculture in Kenya, compiled by R. J. M. Swynnerton, Government Printer, Nairobi, 1954

Development Programme 1954–57, Government Printer, Nairobi, 1955

Colony and Protectorate of Kenya: The Agriculture Ordinance, Ordinance No. 8, 1955

African Land Development in Kenya 1946–1955, Nairobi

Development Programme 1957–60, Government Printer, Nairobi, 1957

Native Lands Registration Ordinance, 1959, No. 27, 1959

CORFIELD, F. D., *The Origins and Growth of Mau Mau, an Historical Survey*, Sessional Paper No. 5, 1959/60, Government Printer, Nairobi, 1960

Report of the Committee on the Organisation of Agriculture, Government Printer, Nairobi, 1960

The Development Programme 1960/63, Sessional Paper No. 4, 1959/60, Government Printer, Nairobi, 1960

Three Year Report, 1958–1960, Ministry of Agriculture, Animal Husbandry and Water Resources, Government Printer, Nairobi, 1961

African Land Development in Kenya, 1946–1962, Ministry of Agriculture, Animal Husbandry and Water Resources, Nairobi, 1962

The Economic Development of Kenya: Report of an Economic Survey Mission, Government Printer, Kenya, 1962

Laws of Kenya: The Agriculture Ordinance (revised edition 1962) cap. 318

Kenya, Central Office of Information Reference Pamphlet 59, H.M.S.O., London, 1963

A National Cash Crops Policy for Kenya, Government Printer, Kenya, 1963

Development Plan for the period from 1 July 1964 to 30 June 1970, Government Printer, Nairobi, 1964

Development Plan for the Period 1965/66 to 1969/70, Republic of Kenya

African Socialism and its Application to Planning in Kenya, Sessional Paper No. 10 of 1965, Government Printer, Nairobi, 1965

Department of Agriculture Annual Reports

Republic of Kenya, the National Assembly, House of Representatives, Official Reports

d. *Unpublished papers*

African Land Utilization and Settlement Board, Quarterly Reports (mimeo), first ending 31 March 1947

African Land Utilization and Settlement Report 1945–1950 (mimeo)

BROWN, R. H. 'A Survey of Grazing Schemes Operating in Kenya' (mimeo), Department of Veterinary Services, 1959

(ii) KENYA: MILLION-ACRE SCHEME

a. *Books*

DE WILDE, JOHN C. *et al.*, *Experiences with Agricultural Development in Tropical Africa*, Vol. 2, 1967, pp. 188–220

HELLEINER, G. K. (ed.), *Agricultural Planning in East Africa*, pp. 118–138, J. D. MacArthur, 'Agricultural Settlement in Kenya', 1968

RUTHENBERG, HANS, *African Agricultural Production Development Policy in Kenya 1952–65*, 1966, pp. 61–86

WHETHAM, Edith H., *Co-operation, Land Reform and Land Settlement*, 1968, pp. 12–18

b. *Monographs and articles*

BELSHAW, D. G. R., 'Agricultural Settlement Schemes in the Kenya Highlands', *East African Geographical Review*, No. 2, 1964, pp. 30–37

BROWN, LESLIE, 'The Settlement Schemes' (a series of weekly articles), *Kenya Weekly News*, 16 July to 22 October 1965

CAREY JONES, N. S., 'The Decolonization of the White Highlands of Kenya', *Geographical Journal*, Vol. 131, Part 2, June 1965, pp. 186–201

CLOUGH, R. H., *Some Economic Aspects of Land Settlement in Kenya: a Report on an Economic Survey in Four Districts of Land Settlement in Western Kenya, 1963–1964*, Egerton College, Kenya, June 1965

— 'Some Notes on a Recent Economic Survey of Land Settlement in Kenya', *East African Economic Review* (New Series), Vol. 1, No. 3, December 1965, pp. 78–83

ETHERINGTON, D. M., 'Land Resettlement in Kenya: Policy and Practice', *East African Economics Review*, Vol. 10, No. 1, June 1963, pp. 22–34

HARBESON, JOHN W., 'Land Resettlement: Problem in Post-Independence Rural Development', *Makerere Institute of Social Research Conference Papers*, No. 407, January 1967

MORGAN, W. T. W., 'The "White Highlands" of Kenya', *Geographical Journal*, Vol. 129, Part 2, June 1963, pp. 140–155

NGUYO, WILSON, 'Some Socio-Economic Aspects of Land Settlement in Kenya', *Makerere Institute of Social Research Conference Papers*, No. 428, January 1967

NOTTIDGE, C. P. R., and GOLDSACK, J. R., *The Million-Acre Settlement Scheme 1962–1966*, Department of Settlement, Nairobi, 1966

STORRAR, A., 'A Guide to the Principles and Practices of Land Settlement in Kenya', *Journal of Local Administration Overseas*, Vol. 3, No. 1, January 1964, pp. 14–19

WHETHAM, EDITH, 'Land Reform and Resettlement in Kenya, *East African Journal of Rural Development*, Vol. 1, No. 1, January 1968, pp. 18–29

c. *Official publications*

Land Tenure and Control Outside the Native Lands, Sessional Paper No. 10, 1958/59, Government Printer, Nairobi, 1959

Department of Settlement, Kenya, Annual Reports 1963–1964, 1964–65, and *1965–66*, Nairobi

d. *Unpublished papers*

Department of Settlement, Departmental Administration Instructions Manual—a Guidebook to Assist Settlement Officers (mimeo)

MAINA, J. W., 'Land Settlement in Kenya: Production under Close Super-vision' (mimeo) Paper to the East African Staff College Seminar on Employment and Agricultural Development, 15 to 29 August 1965

— 'Settlement Patterns and Prospects' (mimeo), paper to Seminar on Rural Development, Syracuse University and University College, Dar es Salaam, 4 to 7 April 1966

ODINGO, RICHARD, 'Land Settlement and Rural Development: Part 2, Land Settlement in the Kenya Highlands', paper to Conference on Education, Employment and Rural Development, University College, Nairobi, 1966

(iii) MWEA IRRIGATION SETTLEMENT

a. *Books*

DE WILDE, JOHN C. *et. al.*, *Experiences with Agricultural Development in Tropical Africa*, Vol. 2, 1967, pp. 221–241

RUTHENBERG, HANS, *African Agricultural Production Development Policy in Kenya 1952–65*, 1966, pp. 58–61

b. *Monographs and articles*

FAIRBURN, W. A., *Geology of the Fort Hall Area*, Geological Survey of Kenya, Report No. 73, Government Printer, Kenya, 1965

GIGLIOLI, E. G., 'Recent Advances in Rice in Kenya—the Mwea Irrigation Settlement', *Agronomie Tropicale*, No. 8, August 1963, pp. 828–833

— 'Mechanical Cultivation of Rice on the Mwea Irrigation Settlement', *East African Agricultural and Forestry Journal*, Vol. 30, No. 3, January 1965, pp. 177–181

— 'Staff Organization and Tenant Discipline on an Irrigated Land Settlement', *East African Agricultural and Forestry Journal*, Vol. 30, No. 3, January 1965, pp. 202–205

— 'Planning of Settlement in Irrigated Areas—Some Aspects of Planning Irrigation Settlement in Kenya', paper S10/67–12 to the Seminar on Problems and Approaches in Planning Agricultural Development, 16 October to 7 November 1967 (U.N.—E.C.A., Addis Ababa and German Foundation to Developing Countries, Berlin), (to be published in collected seminar papers)

LACEY, GERALD, and WATSON, ROBERT, *Report on Rice Production in the East and Central African Territories*, H.M.S.O., London, 1949

MACARTHUR, J. D., 'Labour Costs and Utilization in Rice Production on the Mwea/Tebere Irrigation Scheme', *East African Agricultural and Forestry Journal*, Vol. 33, No. 4, April 1968, pp. 325–334

c. *Official publications*

Department of Agriculture Annual Reports 1953–1964

African Land Development in Kenya 1946–55, Nairobi, pp. 96–97

Upper Tana Catchment Water Resources Survey 1959, report by Sir Alexander Gibb and Partners (Africa), Kenya Government, April 1959

The Native Lands (Irrigation Areas) Rules 1959, Legal Notice 410/59

Kenya Gazette Notice 3099 of 5 July 1960, (setting aside land)

African Land Development in Kenya 1946/1962, Ministry of Agriculture, Animal Husbandry and Water Resources, Nairobi, 1962, pp. 110–112

The Trust Lands (Irrigation Area) Rules 1962, Legal Notice 535/62

Kenya: Report of the Regional Boundaries Commission, Detailed Description of Boundaries, Cmnd. 1899–1, H.M.S.O., London, 1963

Kenya Agricultural Produce Marketing Board First Annual Report, Balance Sheet and Accounts for the financial period ended 31 July 1965, Diamond Press, Nairobi ,1965

The Irrigation Bill, 1965, published in Kenya Gazette Supplement No. 74 (Bills No. 20), Nairobi, 21 September 1965

The Irrigation Act, 1966, No. 13, 1966

d. *Unpublished papers and reports*

MAHER, COLIN, 'Soil Erosion and Land Utilization in the Embu Reserve', (mimeo) (2 vols), 1938

African Land Utilization and Settlement Report 1945–50 (mimeo) (p. 15 —Mwea Development and Reclamation Scheme)

'The Mwea', a report dated 8 January 1953, by a former District Agricultural Officer of Embu District (Ministry of Agriculture file IRRIG/MWEA, folio 15A)

The Irrigation Adviser's report on his visit to Mwea and Tebere Irrigation Projects (typescript), 21 April 1954

BROWN, F. A., 'Review of the Mweia [sic]/Tebere Irrigation Project' (mimeo), Nairobi, 3 January 1955

Managers' Fortnightly, Monthly and Quarterly Reports (variable) 1955–1964

Embu District Gazeteer 1955/56 (mimeo)

Embu District Annual Reports 1957–1962

Annual Reports Mwea/Tebere Irrigation Scheme 1957–1959

Mwea Irrigation Settlement, An Analysis of Rice Yields—1959 and 1960 Short Rain Crops (mimeo)

Mwea Irrigation Settlement, Annual Reports 1960–1967 (mimeo)

Mwea Irrigation Settlement, An Analysis of Yields and Incomes for the 1961 Short Rains Rice Crop (mimeo), also for 1962, 1963, 1964, 1965, 1966 and 1967

'National Irrigation Schemes: Consolidated Capital, Expenditure, and Revenue Figures for the 11 Years to 30 June 1966 (amended)' (mimeo), by the Secretary/Chief Accountant, National Irrigation Board

U.N.I.C.E.F./Makerere Farm Innovation Survey, 1966

National Irrigation Board: General Manager's Annual Report, 1966–1967 (mimeo), Nairobi, National Irrigation Board, Nairobi, 1967

KORTE, ROLF, 'Report on the Nutrition Survey Conducted on the Mwea-Tebere Irrigation Scheme July 1966, March 1967' (mimeo), Medical Faculty, Institute for Human Nutrition, Giessen, 1967

GOLKOWSKY, RUDOLF, 'Structure and Level of Tenant Incomes and Expenses in the Mwea Irrigation Settlement, Kenya' (mimeo), IFO-Institute for Economic Research, Munich, December 1967

(iv) PERKERRA IRRIGATION SCHEME

'Report on Portion of the Kamasia and Njemps Native Reserve with Special Reference to Irrigation Possibilities', report of the Senior Agricultural Chemist, 29 April 1933

CARRICK, H. E., and TETLEY, A. E., 'The Perkerra River Irrigation Project', (mimeo), 25 June 1936

African Land Development in Kenya 1946–1962, Ministry of Agriculture, Animal Husbandry and Water Resources, Nairobi, 1962, pp. 132–135

BROWN, F. A., 'A Review of the Perkerra River Scheme' (mimeo) Nairobi, 20 December 1954

Annual Reports of the Perkerra Irrigation Scheme 1959–1965

District Gazeteer, Baringo District (mimeo), by the District Agricultural Officer, (2nd edition, revised), October 1961

DE WILDE, JOHN C. et al., 'Experiences with Agricultural Development in Tropical Africa', Vol. 2, 1967, pp. 221–241

D. NIGERIA

a. *Books*

BALDWIN, K. D. S., *The Niger Agricultural Project, an Experiment in African Development*, 1957

KREININ, MORDECHAI E., *Israel and Africa, a Study in Technical Co-operation*, 1964, pp. 48–70 (on Western Nigerian Farm Settlements)

WHETHAM, EDITH H., *Co-operation, Land Reform and Land Settlement*, pp. 49–55

b. *Monographs and articles*

AGRAWAL, G. D., *Farm Planning and Management Manual*, Government Printer, Ibadan, 1964

HUNT, E. O. W., *An Experiment in Resettlement*, Government Printer, Kaduna, 1951

KREININ, MORDECHAI E., 'The Introduction of Israel's Land Settlement Plan to Nigeria', *Journal of Farm Economics*, Vol. 45, August 1963, pp. 535–546

NASH, T. A. M., *The Anchau Rural Development and Settlement Scheme*, H.M.S.O., London, 1948

OKEDIJI, OLADEJO O., 'Some Socio-Cultural Problems in the Western Nigeria Land Settlement Scheme: a Case Study', *Nigerian Journal of Economic and Social Studies*, Vol. 7, No. 3, November 1965, pp. 301–310

OLATUNBOSUN, DUPE, 'Nigerian Farm Settlements and School Leavers' Farms—Profitability, Resource Use and Social-Psychological Considerations', *Consortium for the Study of Nigerian Rural Development Report No. 9*, East Lansing, Michigan, Michigan State University, 204 Agricultural Hall, 1968

SCHWAB, WILLIAM B., *An Experiment in Resettlement in Northern Nigeria*, Haverford College, 1954

TAYLOR, C. B., 'An Experiment in Land Settlement', *Tropical Agriculture*, Vol. 20, No. 4, April 1943, pp. 71–73

WELLS, JEROME C., 'The Israeli Moshav in Nigeria: an Estimate of Returns', *Journal of Farm Economics*, Vol. 48, No. 2, May 1966, pp. 279–294

c. *Official publications*

Future Policy of the Ministry of Agriculture and Natural Resources, Sessional Paper No. 9 of 1959 of the Western Regional Legislature, Government Printer, Ibadan, 1959

Land Settlement Scheme—Farm Settlements, Government Printer, Ibadan, June 1960

Western Nigeria House of Assembly Debates Official Report, especially debates of 11 April 1961, 5 April 1962, 12 April 1962, 2 April 1964, 16 April 1964 and 2 April 1965

Eastern Nigeria Development Plan 1962–68, Government Printer, Enugu, 1962

Northern Nigeria Development Plan 1962–68, Government Printer, Kaduna, 1962

Western Nigeria Development Plan 1962–68, Sessional Paper No. 8, 1962, of the Western Nigeria Legislature

Report of the Government of Nigeria on the Farm Settlement Scheme in the Western Region, Report No. 1720, F.A.O., Rome, 1963

Agricultural Development in Nigeria: 1965–1980, FAO, Rome, 1966

Eastern Nigeria Farm Settlement Scheme, Technical Bulletin No. 6, Supplement to Agricultural Bulletin No. 2, Ministry of Agriculture, Enugu, undated

d. *Unpublished papers*

AGRAWAL, G. D., 'A Policy Paper on Farm Settlement Project in Western Region—Nigeria' (mimeo), Ibadan, 1961

EICHER, CARL K., 'Reflections on Capital-Intensive Moshav Farm Settlements in Southern Nigeria' (mimeo), paper presented at ADC seminar on Co-operatives and Quasi-Co-operatives, University of Kentucky, 26 to 30 April 1967

JONES, G, I., (no title, but concerning Farm Settlements in Southern Nigeria) (mimeo), paper to a symposium of the African Studies Association of the United Kingdom, 22 September 1967

KLAYMAN, MAX, 'The Transferability of the Israeli Moshav for the Agricultural Development of Other Countries', paper presented at the ADC Conference on Co-operatives and Quasi-Co-operatives, University of Kentucky, 26 to 30 April 1967

WELLS, JEROME C., 'An Appraisal of Agricultural Investments in the 1962–68 Nigerian Development Program', unpublished Ph.D. Dissertation, Ann Arbor, Michigan, 1962

E. RHODESIA

a. *Books*

BIEBUYCK, DANIEL (ed.), *African Agrarian Systems*, pp. 185–200, C. Kingsley-Garbett, 'The Land Husbandry Act of Southern Rhodesia', 1963

RODER, WOLF, *The Sabi Valley Irrigation Projects*, 1965

YUDELMAN, MONTAGUE, *Africans on the Land*, 1964

b. *Monographs and articles*

ANDREWS, BRUCE, *The Lowveld, an Economic Survey*, Ramsay Parker, Salisbury, 1964

FAIR, T. J. D., 'Rhodesian Lowveld: Source of New Economic Strength', *Optima*, December 1964, pp. 191–201

JOHNSON, R. W. M., 'African Agricultural Development in Southern Rhodesia, 1945–1960', *Food Research Institute Studies*, Vol. 4, No. 2, 1963–1964, pp. 165–223

PENDERED, A., and VON MEMERTY, W., 'The Native Land Husbandry Act of Southern Rhodesia', *Journal of African Administration*, Vol. 7, No. 3, July 1955

POWYS-JONES, L., 'The Native Purchase Areas of Southern Rhodesia', *Journal of African Administration*, Vol. 7, No. 3, 1955

d. *Unpublished papers*

WOODS, ROGER, 'The Dynamics of Land Settlement: Pointers from a Rhodesian Land Settlement Scheme' (mimeo), Dar es Salaam, December 1966

F. SUDAN

a. *Books*

DE SCHLIPPE, PIERRE, *Shifting Cultivation in Africa, the Zande System of Agriculture*, Routledge & Kegan Paul, 1956

GAITSKELL, ARTHUR, *Gezira, a Story of Development in the Sudan*, 1959

REINING, CONRAD C., *The Zande Scheme, an Anthropological Case Study of Economic Development in Africa*, 1966

WHETHAM, EDITH H., *Co-operation, Land Reform and Land Settlement*, 1968, pp. 32–37

b. *Monographs and articles*

ASAD, T., CUNNISON, I., and HILL, L. G., 'Settlement of Nomads in the Sudan: A Critique of Present Plans', *Proceedings of the Conference of the Philosophical Society of the Sudan*, Vol. I, December 1965, pp. 102–125

BEER, G. W., 'The Social and Administrative Effects of Large-Scale Planned Agricultural Development', *Journal of African Administration*, Vol. 5, No. 3, July 1953, pp. 112–118

BRAUSCH, GEORGES, CROOKE, PATRICK, and SHAW, JOHN, *Bashaqra Area Settlements 1963, a Case Study in Village Development in the Gezira Scheme*, University of Khartoum, 1964

BRAUSCH, GEORGES, 'Change and Continuity in the Gezira Region of the Sudan', *International Social Science Bulletin* (Paris), Vol. 16, No. 3, 1964, pp. 340–356

— 'Labour Problems in the Gezira', *Civilizations*, Vol. 13, No. 3–4, 1963

— 'Historical Introduction to a Sociological Analysis of the Gezira', *Sudan Notes and Records*, Vol. 45, 1964

RANDELL, JOHN R., 'El Gedid: a Blue Nile Gezira Village', *Sudan Notes and Records*, Vol. 39, 1958, pp. 25–39

REINING, CONRAD C., 'The History of Policy in the Zande Scheme', *Proceedings of the Minnesota Academy of Science*, Vol. 27, 1959, pp. 6–13

SHAW, D. J., 'Labour Problems in the Gezira Scheme', *Agricultural Economics Bulletin for Africa*, No. 5, April 1964, pp. 1–41

— 'Recent Developments in the Gezira Scheme', *Information on Land Reform, Land Settlement and Co-operatives*, issued by the Food and Agriculture Organization of the United Nations, No. 2, 1964, pp. 8–22

— 'The Managil South Western Extension: an Extension to the Gezira Scheme', *Bulletin No. 9 issued by the International Institute for Land Reclamation and Improvement*, H. Veeman and Zonen N. V., Wageningen, March 1965

SHAW, D. J., 'Resettlement from the Nile in Sudan', *Middle East Journal*, August 1967, pp. 463–487

SIMPSON, S. R., 'Land Tenure Aspects of the Gezira Scheme in the Sudan', *Journal of African Administration*, Vol. 9, No. 2, April 1957, pp. 92–95

THORNTON, D. S., 'The Organization of Production in the Irrigated Areas of the Sudan', *Journal of Agricultural Economics*, Vol. 16, December 1964

— 'Regional Development—The Case of the Northern Province', *Proceedings of the Conference of the Philosophical Society of the Sudan*, Vol. 2, December 1965, pp. 24–47

— 'Contrasting Policies in Irrigation Development, Sudan and India', *University of Reading, Department of Agricultural Economics, Development Studies*, No. 1, September 1966

VERSLUYS, J. D. N., 'The Gezira Scheme in the Sudan and the Russian Kolkhoz: a Comparison of Two Experiments', *Economic Development and Cultural Change*, Vol. 2, No. 1, pp. 32–59

d. *Unpublished papers*

THORNTON, D. S., and WYNN, R. F., 'An Economic Assessment of the Khasm-el-Girba Project' (mimeo), Faculty of Agriculture, University of Khartoum, March 1965

G. TANZANIA

a. *Books*

DE WILDE, JOHN C. *et al.*, *Experiences with Agricultural Development in Tropical Africa*, Vol. 2, pp. 415–450 (Sukumaland)

FUGGLES-COUCHMAN, N. R., *Agricultural Change in Tanganyika, 1945–1960*, Food Research Institute, Stanford University, 1964

LORD, R. F., *Economic Aspects of Mechanized Farming at Nachingwea in the Southern Province of Tanganyika*, H.M.S.O., London, 1963

MALCOLM, DONALD W., *Sukumaland, an African people and their Country, a Study of Land Use in Tanganyika*, Oxford University Press, 1953

NYERERE, JULIUS, *Freedom and Unity: Uhuru na Umoja, a Selection from the Writings and Speeches, 1962-65*, Oxford University Press, 1967

RUTHENBERG, HANS, *Agricultural Development in Tanganyika*, Springer-Verlag, Berlin, 1964

RUTHENBERG, HANS (ed.), *Smallholder Farming and Smallholder Develop-
ment in Tanzania: Ten Case Studies*, especially pp. 249–273 Nikolaus
Newiger, 'Village Settlement Schemes: the Problems of Co-operative
Farming', Weltforum Verlag, Munich, 1968
SMITH, HADLEY E. (ed.), *Agricultural Development in Tanzania*, Institute of
Public Administration Study, No. 2, Oxford University Press, 1965
WOOD, ALAN, *The Groundnut Affair*, 1950

b. *Monographs and articles*

CLIFFE, LIONEL and CUNNINGHAM, GRIFFITHS, 'Ideology, Organisation and
the Settlement Experience in Tanzania' (mimeo), *Rural Development
Paper No. 3*, Dar es Salaam, University College, Rural Development
Research Committee, 1968
COCKING, W. P. and LORD, R. F., 'The Tanganyika Agricultural Corpora-
tion's Farming Settlement Schemes', *Tropical Agriculture*, Vol. 35, 1958
CUNNINGHAM, G. L., 'The Ruvuma Development Association; an Indepen-
dent Critique', *Mbioni*, Vol. 3, No. 2, July 1966, pp. 44–55
GEORGULAS, N., 'Settlement Patterns and Rural Development in Tangan-
yika', *Ekistics*, Vol. 24, No. 141, August 1967, pp. 180–192
IBBOTT, RALPH, 'The Ruvuma Development Association', *Mbioni*, Vol. 3,
No. 2, July 1966, pp. 3–43
LANDELL-MILLS, P. M., 'On the Economic Appraisal of Agricultural
Development Projects: The Tanzania Village Settlement Schemes',
Agricultural Economics Bulletin for Africa, No. 8, December 1966
— 'Village-making in Tanzania', *New Society*, 29 September 1966
NYERERE, JULIUS, 'Socialism and Rural Development', *Mbioni*, Vol. 4,
No. 4, October 1967, pp. 2–46
RWEYEMAMU, ANTHONY H., 'Managing Planned Development: Tanzania's
Experience', *Journal of Modern African Studies*, Vol. 4, No. 1, May
1966
SEGAL, AARON, 'Villages for Tanzania', *Kenya Weekly News*, 12 March
1965
STANNER, W. E. H., 'Sociological Problems of the Groundnut Scheme in
Tanganyika', *Colonial Review*, Vol. 6, No. 2, June 1949, pp. 45–48
THOMAS, GARRY, 'The Transformation Approach at a Tanzania Village
Settlement', *Makerere Institute of Social Research Conference Papers*,
No. 427, January 1967
WOOD, ROBERT E., 'Tanzanian National Institutions, 1: Rural Settlement
Commission', *Mbioni*, Vol. 2, No. 6, November 1965

c. *Official publications*

The Economic Development of Tanganyika, Report of a Mission Organized
by the International Bank for Reconstruction and Development, The
Johns Hopkins Press, Baltimore, 1961
KAPLAN, BENJAMIN, *New Settlements and Agricultural Development in*

Tanganyika: Report and Recommendations to the Government of Tangan-yika Resulting from a Study Mission Sponsored by the Government of Israel, State of Israel, Ministry of Foreign Affairs, Department of International Co-operation, Jerusalem, 1961

President's Address to the National Assembly, 10 December 1962, Government Printer, Dar es Salaam

'Rural Settlement Commission Established', *Press Release No. B/751/ 63LG/5/3*, Tanganyika Information Service, Dar es Salaam, 9 May 1963

YALAN, E., *Report on the Creation of an Organizational Framework for the Villagization of Tanganyika*, State of Israel, Ministry of Foreign Affairs, Department of International Co-operation, Jerusalem, 1963

An Act to Establish a Rural Settlement Commission and for Matters Incidental Thereto, Act No. 62 of 1963, Government Printer, Dar es Salaam, 1964

Rural Settlement Planning, issued by the Rural Settlement Commission, Vice-President's Office, Dar es Salaam, 1964

Tanganyika Five Year Plan for Economic and Social Development 1 July 1964–30 June 1969, Vols. I and II, Government Printer, Dar es Salaam, 1964

The Tanganyika Agricultural Corporation (Amendment) Act, 1964, Government Printer, Dar es Salaam, 1964

President's Address to the National Assembly, 8 June 1965, Mwanainchi Publishing House, Dar es Salaam, 1965

The Rural Settlement Commission (Dissolution) Act of 1965, Government Printer, Dar es Salaam, 1965

President's Speech to the National Assembly on 12 October 1965, Government Printer, Dar es Salaam, 1965

Address by the Second Vice-President, Mr R. M. Kawawa, at the Opening of the Rural Development Planning Seminar at the University College, Dar es Salaam, 4 April 1966, *Press Release IT/I.302*, Ministry of Information and Tourism, Dar es Salaam

Report of the Presidential Special Committee of Enquiry into Co-operative Movement and Marketing Boards, Government Printer, Dar es Salaam, 1966

Proposals of the Tanzania Government on the Recommendation of the Special Presidential Committee of Enquiry into the Co-operative Movement and Marketing Boards, Government Printer, Dar es Salaam, 1966

d. *Unpublished papers*

ALLSEBROOK, G. P., 'The Transformation Approach' (mimeo), Kivukoni College, Dar es Salaam, September 1966

BURKE, FRED, 'Tanzania's Search for a Viable Rural Settlement Policy' (mimeo), paper to a symposium of the African Studies Association of the United Kingdom, 22 September 1967

Colonial Service Second Devonshire Course, 'Rural and Urban Settlement Schemes in Tanganyika and Northern Rhodesia' (mimeo), Cambridge 1952–53

GEORGULAS, NIKOS, 'Technical Problems in Development Administration—the Process of Designing the Structure of Organizations', paper to Seminar on Rural Development, Syracuse University and University College, Dar es Salaam, 4 to 7 April 1966

— 'Structure and Communication: a Study of the Tanganyika Settlement Agency', unpublished D.S.Sc. dissertation, Syracuse University, 1967

LANDELL-MILLS, P. M., 'Village Settlement in Tanzania: An Economic Commentary' (mimeo), University College, Dar es Salaam, November 1965

LEWTON-BRAIN, James, 'The Position of Women on Settlement Schemes in Tanzania' (mimeo), paper to Seminar on Rural Development, Syracuse University and University College, Dar es Salaam, 4 to 7 April 1966

NELLIS, JOHN, 'Planning for Public Support' (mimeo), paper to Seminar on Rural Development, Syracuse University and University College, Dar es Salaam, 4 to 7 April 1966

RIGBY, PETER, 'Settlement Schemes at Kongwa and their Significance for Socio-Economic Development in Ugogo' (typescript), September 1963

RWEYEMAMU, ANTHONY, 'Nation Building and the Planning Process in Tanzania', unpublished Ph.D. thesis, Syracuse University, 1966

SABRY, OSMAN, 'Report on Land Settlement in Tanzania' (mimeo), Dar es Salaam, 1965

SEAL, JOHN B., JR., 'Manpower Utilization Report on the Agricultural Division Ministry of Agriculture, Forests and Wildlife' (mimeo), December 1964

THOMAS, GARRY, 'Effects of New Communities on Rural Areas—The Upper Kitete Example' (mimeo), paper to Seminar on Rural Development, Syracuse University and University College, Dar es Salaam, 4 to 7 April 1966

'Village Settlement in Tanganyika', Memorandum of the Rural Settlement Commission, Division of Settlements, Dar es Salaam, 1963

YEAGER, RODGER, 'Micropolitics, Persistence, and Transformation: Theoretical Notes and the Tanzania Experience' (mimeo), paper to the African Studies Association of the United States, 1967

— 'Micropolitics and Transformation: a Tanzania Study of Political Interaction and Institutionalization', unpublished Ph.D. thesis, Syracuse University, 1968

H. UGANDA

a. *Books*

JACOBS, B. L., *Administrators in East Africa, Six Case Studies*, Government Printer, Entebbe, 1965, pp. 9–51

JOY, J. L. (ed.), *Symposium on Mechanical Cultivation in Uganda*, Uganda Argus for Department of Agriculture, Uganda, 1960

WHETHAM, EDITH H., *Co-operation, Land Reform and Land Settlement*, 1968, pp. 23–31

BIBLIOGRAPHY

b. *Monographs and articles*

BELSHAW, D. G. R., 'An Outline of Resettlement Policy in Uganda, 1945–63', *East African Institute of Social Research Conference Paper*, June 1963

CHARSLEY, SIMON, 'Group Farming in Bunyoro', *East African Institute of Social Research Conference Papers*, January 1966

— 'The Profitability of a Group Farm', *Makerere Institute of Social Research Conference Papers*, No. 405, January 1967

HUTTON, CAROLINE, 'Nyakashaka—A Farm Settlement Scheme in Uganda', *African Affairs*, Vol. 67, No. 267, April 1968, pp. 118–123

ILLINGWORTH, S., 'Preliminary Report on Research: The South Busoga Resettlement Scheme', *East African Institute of Social Research Conference Papers*, January 1963

PURSEGLOVE, J. W., 'Kigezi Resettlement', *The Uganda Journal*, Vol. 14, No. 2, September 1950, pp. 139–152

— 'Resettlement in Kigezi, Uganda', *Journal of African Administration*, Vol. 3, 1951, pp. 13–21

SEGAL, AARON, 'Group Farming in Uganda', *Kenya Weekly News*, 10 July 1964

WATTS, SUSAN J., 'The South Busoga Resettlement Scheme', *Syracuse University Program of Eastern African Studies Occasional Paper No. 17*, Maxwell Graduate School of Citizenship and Public Affairs, Syracuse, April 1966

c. *Official publications*

The Economic Development of Uganda, Report of a Mission Organized by the International Bank for Reconstruction and Development, The Johns Hopkins Press, Baltimore, 1962

Uganda Parliamentary Debates (*Hansard*), especially debates of 26 and 27 June 1963, 6 and 7 July 1964, 16 June 1964, 30 June 1965

The First Five-Year Development Plan 1961/62–1965/66 (sic), Government Printer, Entebbe, 1964

Work for Progress: Uganda's Second Five-Year Plan, 1966–1971, Government Printer, Entebbe, 1966

'The Group Farming Scheme in Uganda', *World Land Reform Conference Paper*, Rome, F.A.O., 1966

Department of Agricultural Special Development Section Annual Report 1967, Namalere Station

The Tractor Hire Service and Group Farm, an Economic Appraisal, Entebbe, Department of Agricultural, Forestry and Co-operatives, 1967

d. *Unpublished papers*

AGRAWAL, G. D., 'Some Considerations Affecting the Organisation of the Mubuku Irrigation Scheme', Rural Development Research Project paper No. 35 (mimeo), Faculty of Agriculture, Makerere University College, 1966

284

CHARSLEY, SIMON, 'Kigumba Settlement: the Establishment of Pluralists' (typescript), 1967
— 'The Group Farm Scheme in Uganda: a Case Study from Bunyoro' (typescript), 1967
HUNT, DIANA M., 'Agricultural Credit in Uganda', unpublished Ph.D. thesis, University of East Africa, 1967
HUTTON, CAROLINE, 'Case Study of Nyakashaka Resettlement Scheme, a preliminary account' (mimeo), East African Institute of Social Research Sociology-Anthropology Workshop, June 1966
— 'Making Modern Farmers' (mimeo), paper to a symposium of the African Studies Association of the United Kingdom, 22 September 1967 (on Nyakashaka)
ILLINGWORTH, SUSAN, 'Resettlement Schemes in Uganda: a Case Study in Agricultural Development', unpublished M.A. thesis, Nottingham University, 1964
KATARIKAWE, E. S., 'Agricultural Aspects of the Kiga Resettlement Programme in Western Uganda', Rural Development Research Project paper no. 11 (mimeo), Faculty of Agriculture, Makerere University College, undated
— 'Some Preliminary Results of a Survey of Kiga Resettlement Schemes in Kigezi, Ankole and Toro Districts, Western Uganda', Rural Development Research Project paper no. 31 (mimeo), Faculty of Agriculture, Makerere University College, undated

I. ZAMBIA

a. *Books*

ALLAN, WILLIAM, *The African Husbandman*, 1965, especially pp. 446–459
BIEBUYCK, DANIEL (ed.), *African Agrarian Systems*, pp. 137–156, Elizabeth Colson, 'Land Rights and Land Use among the Valley Tonga of the Rhodesian Federation: the Background to the Kariba Resettlement Programme', 1963
COLSON, ELIZABETH, *The Social Organization of the Gwembe Tonga*, Human Problems of Kariba, Vol. I, published on behalf of the Rhodes-Livingstone Institute, Northern Rhodesia, Manchester University Press, 1960
KAY, GEORGE, *Changing Patterns of Settlement and Land Use in the Eastern Province of Northern Rhodesia*, University of Hull Occasional Papers in Geography No. 2, University of Hull, 1965
LOWE-MCCONNELL, R. H., *Man-made Lakes*, pp. 99–108, Thayer Scudder, 'Man-made Lakes and Population Resettlement in Africa', 1966
SCUDDER, THAYER, *The Ecology of the Gwembe Tonga*, Kariba Studies Vol. II, published on behalf of the Rhodes-Livingstone Institute, Northern Rhodesia, Manchester University Press, 1962

b. *Monographs and articles*

COLSON, ELIZABETH, 'Social Change and the Gwembe Tonga', *The Rhodes-Livingstone Journal*, No. 35, June 1964, pp. 1–13

JOHNSON, R. W. M., 'The Northern Province Development Scheme, Northern Rhodesia', *Agricultural Economics Bulletin for Africa*, No. 5, April 1954, pp. 42–110 (on Mungwi)

MAKINGS, S. M., 'Agricultural Change in Northern Rhodesia/Zambia, 1945–1965', *Food Research Institute Studies*, Vol. 6, No. 2, 1966, pp. 195–247

REEVE, W. H., 'The Progress and Geographical Significance of the Kariba Dam', *Geographical Journal*, Vol. 126, Part 2, June 1960, pp. 140–146

'The Peasant Farming Scheme in Northern Rhodesia', *Agricultural Economics Bulletin for Africa*, No. 1, pp. 56–57

c. *Official publications*

Report of the Commission appointed to Inquire into the Circumstances Leading up to and Surrounding the Recent Deaths and Injuries Caused by the Use of Firearms in the Gwembe District and Matters Relating Thereto, Government Printer, Lusaka, 1958

d. *Unpublished papers*

Colonial Service Second Devonshire Course, Cambridge 1952–53: 'Rural and Urban Settlement Schemes in Tanganyika and Northern Rhodesia' (mimeo)

Minutes of the Second Meeting of the Land Resettlement Board on 6 March 1967 (mimeo)

'Memorandum for Submission to the Land Resettlement Board Meeting on March 6, 1967' (mimeo)

'Land Resettlement Board, Chairman's Speech' (mimeo), March 1967

'Land Resettlement Board, Minister's Address' (mimeo), March 1967

J. BIBLIOGRAPHIES AND REFERENCE

Africa Research Bulletin, Political, Social and Cultural Series

Africa Research Bulletin, Economic, Financial and Technical Series

BROWN, EDWARD, and WEBSTER, JOHN (compilers), *A Bibliography on Politics and Government in Kenya*, Program of Eastern African Studies Occasional Bibliography No. 1, Maxwell Graduate School of Citizenship and Public Affairs, Syracuse, September 1965

C.S.A./C.C.T.A., *Inventory of Economic Studies Concerning Africa South of the Sahara*, 1963

Department of State, Agency for International Development: *Development Administration and Assistance, an Annotated Bibliography*, Washington, July 1963

Department of Technical Co-operation: *Public Administration, a Select Bibliography*, London, 1963

FAO: *Catalogue of FAO Publications 1945–64*, Food and Agriculture Organization of the United Nations, Rome, 1965

GARLING, ANTHEA, *Bibliography of African Bibliographies*, Occasional Papers No. 1, African Studies Centre, Cambridge, 1968

HAZELWOOD, ARTHUR, *The Economics of 'Under-Developed' Areas, an Annotated Reading List of Books, Articles, and Official Publications*, 2nd enlarged edition, Oxford University Press, 1962

Library of Congress, *Agricultural Development Schemes in Sub-Saharan Africa, a Bibliography*, compiled by Ruth S. Freitag under the direction of Conrad C. Reining and Walter W. Deshler, Library of Congress, General Reference and Bibliography Department, Washington, 1963

Ministry of Overseas Development, *Public Administration, a Select Bibliography*, 1st supplement, London, 1964

MOLNOS, ANGELA, *Die Sozialwissenschaftliche Erforschung Ostafrikas 1954–1963*, Afrika-Studien Nr. 5, IFO-Institut fur Wirtschaftsforschung Afrika-Studienstelle, Springer-Verlag, Berlin, 1965

Nigerian Institute of Social and Economic Research, *Research for National Development, a Survey of Recent, Current and Proposed Research on Problems Relating to Nigerian Economic and Social Development* (mimeo), University of Ibadan, April 1966

SPITZ, ALAN A., and WEIDNER, EDWARD W., *Development Administration, an Annotated Bibliography*, East-West Center Press, Honolulu, 1963

Standing Conference on Library Materials on Africa, *United Kingdom Publications and Theses on Africa, 1963*, Heffer, Cambridge, 1966

Standing Conference on Library Materials on Africa, *United Kingdom Publications and Theses on Africa, 1964*, Heffer, Cambridge, 1966

Index

A.A.O., *see* Agricultural Officer, Assistant
Accra, 206, 210–12
administrative officers, 16, 17
'Administrators', 213, 214, 217
Administration, the, 16–19, 21, 47–54, 56–7, 59–74, 80–111, 122, 131, 140, 195, 197, 211–14
  See District Commissioner, District Officer, Provincial Commissioner
aerial spraying, 87
African Court, Wanguru, 121
African District Councils (A.D.C.s), 51, 128
African Land Development Organization, 30, 39, 49, 52–63, 68, 69, 142, 190, 195–8, 208, 209, 227, 232
African Staff Committee, 119
Africanization, 107, 108, 111, 140, 141, 150, 169
Agricolas, xvi–xvii, 47, 48, 52–8, 68–80, 88–92, 97, 100, 102, 128, 132, 208, 211–17
Agricultural Committee, District, 109, 111
  Officer, Assistant, 85, 107, 118, 120, 136, 164, 166–9, 172, 180, 257
    District, 52, 56, 218
    Provincial, 52, 99, 114, 127
  Research Officer, 72, 74, 114
Agriculture. Department(s) of, 49, 50, 52, 54, 68, 100, 141, 142, 192, 195, 196, 215–18
  Director of, 53, 62, 63, 64, 85, 98
  Ministry of, 83, 85, 97, 108, 109, 111, 123, 127, 145, 146, 218
  Secretary for, 98
Agriculturalist, Chief, 109, 123
Ahero Pilot Irrigation Project, 136, 232
ahoi, 47, 48, 68, 71, 81, 86, 101, 102, 191
Akin Deko, Chief, 258
ALDEV, *see* African Land Development Organization
Alvord, Simon, 144, 258
Amani, 153
Apthorpe, Raymond, 11, 34, 228, 260
Arabic, 168
Anchau Rural Development and

Settlement Scheme, 22, 151, 232
Aswan dam, 151, 258
Azande, 28, 29, 155, 157, 159, 161, 171, 183

Bantu, 44
Baldwin, K. D. S., 8, 13, 162, 213
Baringo, 90
Belshaw, D. G. R., 11, 155, 199, 253
bilharzia control measures, 192, 198
Blatte Valley, 6
Block Inspector, 94
bonus schemes, 126
Botswana, 6
Bridger, G. A., 10
British colonial rule, impact of, 16, 17, 18
  Volunteer Programme, 143, 145
Brown, Leslie, 253, 256
Bunyoro Agricultural Company, 29
Bushubi, 181

camps, 61, 64, 72, 86
Canals, construction of, 59, 61, 62, 77, 78, 181, 209, 210
Carr, Stephen, 33
cash economy, 18
cash crop, 3, 10, 30, 31, 131, 173, 180
casual labour, 145
cattle, 22, 53, 113
Charsley, Simon, 167, 185
Chesa Scheme, 34, 161, 175–7, 228, 232, 251, 261
Chiefs, 50, 51, 152
Christian missions, 18
Clerks, 145, 149
Clough, R. H., 253
cocoa, 238
Cohen, Andrew, 26
Colonial Development Act, 1929, 19
  and Welfare Act, 1940, 23
  Corporation, 28, 213
  Fund, 19
  Primary Products Committee, 1947, 27
Commissioner, Chief Native, 49, 53
communications, 77
community development, 61
Co-operation, Departments of, 195, 218
Co-operative Officer, District, 218
co-operatives, 3, 25–7, 31, 34, 38, 113, 127–8, 134, 184, 215–18, 228

289